Montague Chamberlain

A Catalogue of Canadian Birds

With Notes on the Distribution of the Species

Montague Chamberlain

A Catalogue of Canadian Birds
With Notes on the Distribution of the Species

ISBN/EAN: 9783337204617

Printed in Europe, USA, Canada, Australia, Japan

Cover: Foto ©berggeist007 / pixelio.de

More available books at **www.hansebooks.com**

CANADIAN BIRDS.

A CATALOGUE

OF

CANADIAN BIRDS,

WITH

NOTES ON THE DISTRIBUTION OF THE SPECIES.

BY

MONTAGUE CHAMBERLAIN.

SAINT JOHN, N. B.
J. & A. McMILLAN, 98 PRINCE WILLIAM STREET.
1887.

Entered, According to Act of Parliament of Canada, in the year 1887,
BY MONTAGUE CHAMBERLAIN,
In the Office of the Minister of Agriculture, at Ottawa.

PREFACE.

The object of this Catalogue is to bring together the names of all the birds that have been discovered within the boundaries of the Dominion, from the Atlantic Ocean to the Pacific, and north to the Arctic; to present these in the system of nomenclature and in the sequence now generally adopted by American Ornithologists, and to give the geographical distribution of each species.

This latter portion of the work has not been accomplished very satisfactorily, for, although considerable labor and care have been devoted to the preparation of the notes on distribution, they are not at all complete, and I fear that, on further investigation, some of them will be proven incorrect. There is no way at present of avoiding these defects. All the information that is now obtainable has been procured; the works of the older authors have been freely drawn upon, as well as those of recent writers, and a number of MSS. reports have been prepared expressly for the present work. But the greater portion of the country—immense stretches of forest and prairie and sea coast—have received little attention from Ornithologists, while even the more settled districts have not yet been fully investigated, leaving a large amount of field-work still to be done before anything like a complete account of the Birds of Canada can be produced.

I am quite aware that this opinion regarding the narrow limits of our knowledge of Canadian birds is opposed to that held by some of the leading scientific men of the Dominion, who consider that all that can be learned about our fauna is now known to science. That these gentlemen have held such a view has been unfortunate for Canadian ornithology, for it has led them to take little interest in the investigations that were being made, and to withhold all encouragement from students of the science; and as this indifference of our leaders in scientific affairs has been, it appears to me, the one great reason why ornithology has not made the advance in Canada that it has in other countries, I take this opportunity of drawing attention to the matter.

Lest the reader might think that my idea of the extent and importance of the work yet required is exaggerated, I will quote some extracts bearing on this subject from letters of prominent Naturalists who have made a special study of American birds. These letters were addressed to me privately, and were not written for publication, but as the matter is exceedingly instructive, coming from such eminent authorities, I have asked permission to print it.

The following is from the pen of Prof. J. A. ALLEN, of the American Museum of Natural History, New York, who has been President of the American Ornithologists' Union ever since its formation, and who takes rank among the foremost of American Zoologists. Prof. ALLEN writes:

"I have long watched with interest the reports of the Canadian Survey, and have been disappointed to find the Natural History portion of the work receiving so small a share of attention, where the field is so inviting and as yet so little worked. The birds and mammals of British North America offer a particularly attractive field for research. While we know in a general way what species occur there, and somewhat of their distribution, many problems of exceeding interest in relation to North American birds and mammals can be settled satisfactorily only by means of extensive field-work and large series of specimens gathered in the great regions north of the United States. It is in this vast territory we are to look for many of the connecting links between various northern forms of birds and mammals. In respect to the latter we are especially lacking in material, the want of which seriously interferes with intelligent work. Doubtless not a few new species, and a considerable number of new sub-species, await discovery in Canadian territory; while our knowledge of the manner of occurrence and distribution of the birds and mammals generally in this region is extremely unsatisfactory. No portion of this continent north of Mexico offers so inviting a region for natural history exploration as the great northern interior, where only the most superficial harvest has been reaped."

Dr. ELLIOTT COUES, of Washington, whose brilliant scholarship has so enriched the literature of American ornithology, and who is the best known in Canada of recent authorities, writes to me thus:

Preface.

"As you are aware there has of late been a good deal of discussion, here and among the leaders of the American Ornithologists' Union, respecting the comparative status of Canadian ornithology, I am tempted to write to you, as our leading Canadian member, and trust you will not take it amiss if I call your attention to the great amount of work that needs to be done before your country can stand side by side with the United States in this branch of science. With the exception of Mr. McIlwraith's work—the best manual we have on the special subject—most of the recent advances are due to the Geological Survey, directly or indirectly. And this leads me to inquire whether it would be possible for the Survey to undertake the requisite work in a more systematic manner, even to the extent of including some professed Ornithologist in its corps. * * * * "

Mr. Robert Ridgway, the well known Curator of the Department of Birds at the National Museum, Washington, and who is the author of several of the leading standard works on American ornithology, writes:

"I trust the effort to create an interest in ornithology among Canadian students will, as it certainly should, prove successful.

"There are so many interesting, and, from a scientific standpoint, important problems regarding the distribution (both general and local), migrations, etc., of our birds yet to be worked out, that you will have the best wishes of all naturalists for your success.

"Much as has been done toward the development of ornithology in North America, it can be safely said that we know the subject only in outline; and I do not think I exaggerate when I say that less is known of the birds of the Provinces than of those of any equal area in the 'far-western' portions of the United States, for the latter have become so well explored by our numerous Government Surveys."

In a letter received from Dr. C. Hart Merriam, Chief of the Division of Ornithology and Mammalogy of the Bureau of Agriculture, Washington, and who is well known in Canadian scientific circles, there occurs the following reference to the subject:

"I am glad to learn from your recent letter that you are endeavoring to create an increased interest in ornithology among Canadian students, and hope you will succeed. * * * * * * * *

"The Geological Survey has done a vast amount of splendid work in botany. * * * Why should it not do equally good work in ornithology? Surely the economic importance of the subject would justify many times the expenditure."

WILLIAM BREWSTER, Esq., who is in charge of the Department of Birds and Mammals at the Museum of Comparative Zoology, Cambridge, and the Boston Society of Natural History, asks:

"What has Canada done for ornithology? In general terms, simply nothing, excepting the little that has resulted from purely private investigations, or from work instigated, and in some cases paid for, on this side of the line. The results of this work are trifling compared to the as yet untried opportunities. * * * Speaking in general terms, Canada — and especially its Northwestern Provinces — is still a virgin field, about which we are in almost total ignorance. The opportunity it offers is surely tempting."

I have heard Dr. SCLATER, Secretary of the Zoological Society of London, express similar views regarding the importance of the ornithological work still to be done within the boundaries of the Dominion; and I gather from letters received from Sir WILLIAM DAWSON that he is not among those who consider Canadian ornithology a finished work.

But besides the defects arising from insufficiency of material, the present notes on distribution will probably be found to contain mistakes due to the limited experience of some of the younger observers whose records are quoted, though with the care that has been taken to have these verified, they should be few.

I trust, however, that with all its faults and errors, the work may prove of some advantage to all who take an interest in our birds, and of some service to my fellow students. If it assists the latter in their present studies, and incites them to increased activity, the labor will not have been in vain; and I leave the matter here, with the hope that at some day in the near future sufficient material may have been gathered from which a more extended and satisfactory account of Canadian birds can be produced.

The system of nomenclature and classification adopted is that prepared by the Committee of the American Ornithologists' Union, and published in the *Code of Nomenclature and Check-List of North American Birds*, issued by the A. O. U. I have included the European

House Sparrow (*Passer domesticus*), which was omitted from the *A. O. U. Check-List*, and have noted in the Appendix the additions and the changes in nomenclature that have been made since the *Check-List* was issued.

This being but a preliminary work, for the sake of brevity and simplification the synonyms and other technicalities and details are omitted.

It but remains for me to express here the thanks I owe for MSS. reports, and other friendly assistance in the preparation of this work, to Mr. ERNEST E. THOMPSON; Mr. JOHN FANNIN; THOMAS MCILWRAITH, Esq.; J. MATTHEW JONES, Esq.; JAMES M. LEMOINE, Esq.; JOHN NEILSON, Esq.; Mr. JOSEPH M. MACOUN; Mr. WILLIAM E. SAUNDERS; WILLIAM BREWSTER, Esq.; The Rev. VINCENT CLEMENTI; Dr. J. H. GARNIER; Prof. A. H. MCKAY; Dr. G. A. MCCALLUM; Mr. JONATHAN DWIGHT, JR.; Mr. FRANCIS BAIN; Mr. JAMES MCKINLEY; Mr. W. W. DUNLOP; Mr. ERNEST D. WINTLE; Mr. NAPOLEON A. COMEAU; Mr. PHILIP COX, Jr.; Mr. JOHN BRITTAIN; Mr. GEORGE A. BOARDMAN; Mr. LOUIS M. TODD; Mr. HOWARD H. MCADAM; Mr. WILLIAM L. SCOTT; Mr. JOHN A. MORDEN; Mr. R. B. SCRIVEN; Mr. D. KEUTZING; Mr. JAMES W. BANKS; The Rev. DUNCAN ANDERSON; Mr. JOHN G. EWART; Dr. C. K. CLARKE; Mr. WILLIAM L. KELLS; Mr. W. A. SCHOENAN; Mr. WILLIAM YATES; Mr. JOHN B. SPURR; Mr. GEORGE SOOTHERAN, and Mr. CHARLES CLARKE.

M. CHAMBERLAIN.

St. John, N. B.,
  December, 1887.

CATALOGUE.

CATALOGUE.

Æchmophorus occidentalis.
WESTERN GREBE.

This, as its name denotes, is a bird of the west. It is found in abundance on the coasts of British Columbia, and in Manitoba; but east of the latter Province it occurs only as an accidental straggler. The OTTAWA CLUB report that two specimens have been shot at the mouth of North Nation River, and Mr. WM. COWPER has purchased one in the Montreal market.

Colymbus holbœllii.
HOLBŒLL'S GREBE.

This species was formerly called the "Red-Necked Grebe." It occurs throughout the Dominion, from the southern boundary to the Fur Countries, but is rather rare, excepting in a few localities. It is a common resident of British Columbia, and is common during the breeding season in Northwestern Manitoba, and is not uncommon at the mouth of the Bay of Fundy during the migrations.

Colymbus auritus.
HORNED GREBE.

This is the "Spirit-Duck" or "Hell-Diver" of the country sportsmen. It is a common bird throughout the Dominion, breeding from about latitude 45° northward to the higher Fur Countries.

Colymbus nigricollis californicus.
AMERICAN EARED GREBE.

This species is common from Manitoba to the Pacific coast, breeding near all the deep pools of the plains, as well as on the margins of the larger lakes in the mountain districts. Dr. GARNIER reports having taken one at Lucknow, Ontario.

Podilymbus podiceps.
PIED-BILLED GREBE.

This is the "Dab-chick" of the older authors. It is found throughout Canada, north to York Factory and Great Slave Lake, breeding in all suitable localities.

Urinator imber.
LOON.

An abundant bird all over the Dominion, breeding from the southern boundary northward. It winters as far south as the Gulf of Mexico.

Urinator adamsii.
YELLOW-BILLED LOON.

A northern bird, that has not been taken south of the 60th parallel, nor east of Hudson's Bay.

Urinator arcticus.
BLACK-THROATED LOON.

This bird is reported as being "not uncommon" along the coast of British Columbia (*Fannin*), but to the eastward of that Province it is of rare or casual occurrence. A few specimens have been observed in Lake Superior, Lake Erie, and Lake Misstassini, as well as in the Bay of Fundy; it is also sometimes seen in the Hudson's Bay region, but TURNER did not meet with it in the Ungava district, and KUMLIEN reports seeing a few examples only in Cumberland Sound and Grinnell Bay.

Urinator pacificus.
PACIFIC LOON.

A few examples of this species have been taken in the Straits of Fuca, B. C., and it has also been reported as occurring in Great Slave Lake.

Urinator lumme.
RED-THROATED LOON.

This species, the "Red-Throated Diver" of the older authors, occurs throughout the Dominion. It is a migrant only in the more southern localities, but breeds from about the 45th parallel to the Arctics.

Lunda cirrhata.
TUFTED PUFFIN.

This species occurs on the Pacific coast, from California to Alaska, and a few stragglers are found in the Bay of Fundy (*Boardman.*)

Fratercula arctica.
PUFFIN.

An abundant bird along the Atlantic coast, breeding from the Bay of Fundy northward. A few examples have been taken in Lake Superior.

Fratercula arctica glacialis.
LARGE-BILLED PUFFIN.

A northern form of the common Puffin; it occurs very rarely south of Baffin's Bay, though AUDUBON reported finding one specimen at Grand Manan, in the Bay of Fundy.

Fratercula corniculata.
HORNED PUFFIN.
This species is quite common on the west coast of Vancouver and among the Queen Charlotte Islands (*Fannin*).

Cerorhinca monocerata.
RHINOCEROS AUKLET.
A winter visitor to the west coast of Vancouver Island.

Ptychoramphus aleuticus.
CASSIN'S AUKLET.
A common winter visitor to the west coast of Vancouver Island.

Synthliboramphus wumizusume.
TEMMINCK'S MURRELET.
This species occurs occasionally, in winter, along the coast of British Columbia.

Brachyramphus marmoratus.
MARBLED MURRELET.
A very abundant resident of all the coast waters of British Columbia, from Fraser River to Alaska.

Cepphus grylle.
BLACK GUILLEMOT.
An abundant bird on the Atlantic coast, where it is commonly called the "Sea Pigeon"; breeding from the Bay of Fundy to Labrador. It has also been taken, occasionally, in Lake Superior and Lake Erie, and at Hamilton.

Cepphus Mandtii.
MANDT'S GUILLEMOT.
An Arctic species, breeding as far south as Hudson's Bay and Labrador.

Cepphus columba.
PIGEON GUILLEMOT.
An abundant resident of the entire coast line of British Columbia, from Fraser River northward.

Uria troile.
MURRE.
This species was formerly called the "Common Guillemot." It is abundant in the Gulf of St. Lawrence and Labrador, and occurs occasionally along the Atlantic coast of Nova Scotia, and in the Bay of Fundy. A few have been taken in the Great Lakes.

Uria troile californica.
CALIFORNIA MURRE.
This variety occurs on the west coast of Vancouver and among the Queen Charlotte Islands.

Uria lomvia.
BRÜNNICH'S MURRE.
BREWSTER reports this species breeding abundantly on the Magdalene Islands; WELCH found it on the Newfoundland coast; and TURNER reports it at Hudson's Straits. It is abundant in the Bay of Fundy in winter, and BOARDMAN thinks a few breed there. It occasionally straggles as far west as Lake Ontario.

PALLAS'S MURRE (*uria lomvia arra*) occurs in the North Pacific, but there is no record of any examples having been taken in Canadian waters.

Alca torda.
RAZOR-BILLED AUK.

This bird is called "Tinker" by the Gulf of St. Lawrence fishermen, with whom it is familiar. It breeds abundantly on some of the islands in the Gulf, as well as along the Labrador coast. It is common in the Bay of Fundy in winter, and a few individuals may breed there. Several specimens have been taken in Lake Superior.

Plautus impennis.
GREAT AUK.

This species is probably extinct. Portions of skeletons have been found on the shores of the Bay of Fundy, and along the coast of Labrador.

Alle alle.
DOVEKIE.

This is the "Sea Dove" of the older books. It breeds in great abundance from Northern Labrador to the Arctic Ocean, but south of Hudson's Straits occurs only as a winter visitor.

Megalestris skua.
SKUA.

A North Atlantic species, very seldom seen south of Hudson's Straits.

Stercorarius pomarinus.
POMARINE JAEGER.

An Arctic bird. In winter it appears along the coast of Nova Scotia, and in the Bay of Fundy, and has been taken in Lake Ontario and Lake Erie.

Stercorarius parasiticus.
PARASITIC JAEGER.

This species was formerly called Richardson's Jaeger. It summers in the Arctics, and during winter goes as far south as Brazil and Chili, occurring as a regular visitant to the Bay of Fundy in its migrations.

Stercorarius longicaudus.
LONG-TAILED JAEGER.

This species is quite common in the Bay of Fundy during its migrations, though reported as "rare" between that locality and the Arctics. A few have been taken in Hudson's Bay.

Gavia alba.
IVORY GULL.

An Arctic species that has been seldom seen in the temperate latitudes of Canada. A few examples have straggled, accidentally, to the Bay of Fundy and to Lake Ontario.

Rissa tridactyla.
KITTIWAKE.

It is rather misleading to give the habitat of this species as "Arctic regions," for, though KUMLIEN did find it abundant in Cumberland Bay and at Disko, and Dr. BELL found it "especially numerous at Cape Chudleigh," it has been found breeding in numbers on the islands of the Gulf of St. Lawrence by VERRILL, CORY, and BREWSTER, and COMEAU reports its breeding near Point des Monts; while it is known to breed; also, along the Atlantic coast of Nova Scotia and in the Bay of Fundy.

It has been taken during winter in the vicinity of Montreal, and in the Great Lakes.

Rissa tridactyla pollicaris.
PACIFIC KITTIWAKE.

This is the form of the Kittiwake that occurs on the west coast of Vancouver Island (*Fannin*), and north to Bering's Sea.

Larus glaucus.
GLAUCOUS GULL.

The older authors called this bird the "Burgomaster." It is an Arctic species, but a few examples are found in winter along the Atlantic coast of the Maritime Provinces, and in the Great Lakes. It is found, also, in winter, on the west coast of Vancouver Island.

Larus leucopterus.
ICELAND GULL.

This was formerly known as the "White-winged" Gull. It is an Arctic species, and is found, in winter only, along the Atlantic coast of the Maritime Provinces, straying occasionally to the Great Lakes.

Larus glaucescens.
GLAUCOUS-WINGED GULL.

A Pacific species that occurs along the coast from "Alaska south to California." Mr. FANNIN has observed it only on the north-west shore of Vancouver, but Dr. DAWSON took it at Cullen Harbour, Queen Charlotte Sound (*Whiteaves*).

Larus kumlieni.
KUMLIEN'S GULL.

A few examples of this lately described Gull have been taken in the Bay of Fundy in winter. KUMLIEN found it breeding in Cumberland Bay, where it is quite common.

Larus marinus.
GREAT BLACK-BACKED GULL.

The range of this species extends from Cumberland Bay to Long Island (in winter). Not many years ago large numbers built their nests on the islands at the mouth of the Bay of Fundy, but now very few are found there at that season, and between the egg-hunter and the skin-dealer must rest the blame for having driven them to seek more remote and less accessible breeding grounds. During the winter months a number are seen in the Bay of Fundy, and a few are found during the coldest months in the vicinity of Montreal, and in the Great Lakes.

Larus occidentalis.
WESTERN GULL.

A common bird on the coast of British Columbia.

Larus argentatus.
HERRING GULL.

A straggler from Europe that is occasionally met with along our North Atlantic coast.

Larus argentatus smithsonianus.
AMERICAN HERRING GULL.

The common gull of our harbours and lakes; breeding in abundance in suitable localities throughout the Dominion. In winter it ranges as far south as Cuba, though at that season it is common as far north as the 45th parallel.

Larus cachinnans.
PALLAS'S GULL.

Occurs on the west coast of Vancouver Island and northward.

B

Larus californicus.
CALIFORNIA GULL.

Occurs on the west coast of Vancouver Island and northward.

Larus delawarensis.
RING-BILLED GULL.

This species is abundant in the larger lakes of the interior, especially the saline lakes of the Great Plains, but is rare along the sea coasts.

Larus brachyrhynchus.
SHORT-BILLED GULL.

Occurs on the west coast of Vancouver Island and among the Queen Charlotte Islands, and has been taken, also, in Hudson's Bay.

Larus canus.
MEW GULL.

A bird of the Eastern Hemisphere that has occasionally been met with on the Labrador coast.

Larus heermanni.
HEERMANN'S GULL.

A bird of the Pacific; occurs from the Strait of Georgia northward.

Larus atricilla.
LAUGHING GULL.

A southern species, that is found occasionally in the Bay of Fundy. Dr. MORRIS GIBBS gives it as "very abundant" in his

Birds of Michigan, though Mr. McIlwraith has no record of its occurrence in Lake Ontario or Lake Erie. Dr. Hall considered it rare near Montreal, and only one specimen has been reported from Ottawa; Dr. Bell reports it occurring in Hudson's Bay.

Larus franklinii.
FRANKLIN'S GULL.

This species is confined chiefly to the Lakes between Manitoba and the Rockies, breeding from Lake Winnipeg northward. A few examples have wandered as far East as Lake Ontario.

Larus philadelphia.
BONAPARTE'S GULL.

An abundant bird throughout the entire Dominion, breeding from about the 45th parallel northward.

Rhodostethia rosea.
ROSS'S GULL.

Occurs in the Arctic regions.

Xema sabinii.
SABINE'S GULL.

This species has, occasionally, been observed about the mouth of the Bay of Fundy during the winter, though it seldom wanders south of the Arctic circle. Kumlien saw two examples only during his visit to Cumberland Bay, etc., and Turner met with but one at Ungava; one has been taken in Lake Erie (*Wheaton*), and one in Lake Michigan (*Nelson*). Dr. Bell reports having observed it at Port Burwell in September.

Gelochelidon nilotica.
GULL-BILLED TERN.

Although said to be "nearly cosmopolitan," this is merely an accidental visitant to the waters of Canada. One specimen was taken in the Bay of Fundy, in August, 1879, and a few have been captured in the Great Lakes.

Sterna tschegrava.
CASPIAN TERN.

During the spring and fall migrations a few examples of this species have been observed along the Atlantic coast and in the Great Lakes.

Sterna maxima.
ROYAL TERN.

A few examples of this species have been taken in the Great Lakes.

Sterna sandvicensis acuflavida.
CABOT'S TERN.

Three examples of this species were seen near Lucknow, Ontario, by Dr. GARNIER, in the autumn of 1881, and one of them was secured. Its usual habitat is the tropical portions of America.

Sterna forsteri.
FORSTER'S TERN.

This species breeds abundantly in the large lakes of Manitoba, and a few breed regularly on the St. Clair Flats, Ontario. Eastward of that locality it has been reported only at Lake Mistassini (*Macoun*) and Prince Edward Island (*Cory*).

Sterna hirundo.
COMMON TERN.

An abundant summer resident from Lake Manitoba to the Atlantic, and from the Great Lakes to the Fur Countries. Winters from Virginia south.

Sterna paradisæa.
ARCTIC TERN.

Occurs throughout the entire Dominion, but is most abundant on the Atlantic coast, and rather rare in the interior.

Sterna dougalli.
ROSEATE TERN.

A southern bird that sometimes wanders up to the Great Lakes, and has been seen occasionally near the mouth of the Bay of Fundy. One was taken near Halifax in 1868.

Sterna antillarum.
LEAST TERN.

A few examples of this species have been observed in Lake Ontario, and along the Atlantic coast as far north as Labrador, but these were merely accidental stragglers from the south.

Hydrochelidon nigra surinamensis.
BLACK TERN.

This species occurs regularly throughout the Dominion, from Montreal to the Pacific coast, being most abundant in Western Ontario and Manitoba. The only Atlantic coast record is of three specimens taken at Grand Manan, in August, 1879 (*Ruthven Deane*).

Hydrochelidon leucoptera.
WHITE-WINGED BLACK TERN.

Professor MACOUN reports having seen six examples of this species in Swan River, about two degrees north of Winnipeg, on September 1, 1881. Its usual habitat is the Eastern Hemisphere, though it has been taken in Wisconsin (*Kumlien*).

Rynchops nigra.
BLACK SKIMMER.

An accidental straggler to the Bay of Fundy from the south.

Diomedea nigripes.
BLACK-FOOTED ALBATROSS.

This species appears occasionally along the coast of British Columbia.

Diomedea albatrus.
SHORT-TAILED ALBATROSS.

A few examples of this species have been observed along the Pacific coast.

Thalassogeron culminatus.
YELLOW-NOSED ALBATROSS.

One example of this species was captured in the Gulf of St. Lawrence, in September, 1885, and the skin is preserved in the Museum of Laval University, Quebec.

Phœbetria fuliginosa.
SOOTY ALBATROSS.

This species occurs on the British Columbia coast (*Fannin*).

Fulmarus glacialis.
FULMAR.

An Arctic species that occurs as far south as the Straits of Belle Isle, and, accidentally, in the Bay of Fundy.

Fulmarus glacialis minor.
LESSER FULMAR.

A North Atlantic variety.

Fulmarus glacialis glupischa.
PACIFIC FULMAR.

The western form of the common Fulmar. It occurs on the west coast of Vancouver Island.

Puffinus major.
GREATER SHEARWATER.

This species is met with in summer on the Atlantic, from the Bay of Fundy to Greenland.

Puffinus puffinus.
MANX SHEARWATER.

An accidental straggler to the Nova Scotia coast from the eastern side of the Atlantic.

Puffinus stricklandi.
SOOTY SHEARWATER.

This is the "Black Hagdon" of the fisherman. KUMLIEN found it common from the Straits of Belle Isle to Grinnell Bay, and it is occasionally seen off the Nova Scotia coast as far south as the mouth of the Bay of Fundy.

Procellaria pelagica.
STORMY PETREL.

This is rarely seen very near to our shores, but Mr. CORY found it quite common a short distance at sea off the Gulf of St. Lawrence. It is occasionally observed off Labrador and Nova Scotia, and one example has been taken at the mouth of the Bay of Fundy.

Oceanodroma furcata.
FORK-TAILED PETREL.

A Pacific bird, common off the west coast of Vancouver Island.

Oceanodroma leucorhoa.
LEACH'S PETREL.

An abundant bird on both the Atlantic and Pacific coasts.

Oceanites oceanicus.
WILSON'S PETREL.

Occurs on the Atlantic coast, from the Bay of Fundy to Northern Labrador.

Phaëthon flavirostris.
YELLOW-BILLED TROPIC BIRD.

An example of this species, taken at Shubenacadie, N. S., September 6, 1870, is preserved in the Halifax Museum.

Phaëthon æthereus.
RED-BILLED TROPIC BIRD.
One example of this species has been taken on the Newfoundland Banks.

Sula bassana.
GANNET.
This species is very abundant in the Gulf of St. Lawrence and along the Atlantic coast of Labrador, and specimens are seen occasionally in the Great Lakes. It winters in the Gulf of Mexico.

Phalacrocorax carbo.
CORMORANT.
This species is rather rare along the more southern portions of our Atlantic coast, but is not uncommon on the islands in the Gulf of St. Lawrence, and on Mecattina Island, off Labrador. A few specimens have been taken in Lake Ontario.

Phalacrocorax dilophus.
DOUBLE-CRESTED CORMORANT.
An abundant bird in the Maritime Provinces and Labrador; also, a common summer resident in Manitoba. It occurs rather sparingly, as a migrant, in the Great Lakes.

Phalacrocorax dilophus floridanus.
FLORIDA CORMORANT.
A few examples of this southern race have been taken in Lake Erie.

C

Phalacrocorax dilophus cincinatus.
WHITE-CRESTED CORMORANT.

A rather common resident of the coast of British Columbia.

Phalacrocorax penicillatus.
BRANDT'S CORMORANT.

A rather common resident of the coast of British Columbia (*Fannin*).

Phalacrocorax pelagicus robustus.
VIOLET-GREEN CORMORANT.

This species occurs in abundance along the shores of Vancouver Island (*Fannin*).

Phalacrocorax pelagicus resplendens.
BAIRD'S CORMORANT.

Fairly common around Vancouver Island.

Pelecanus erythrorhynchos.
AMERICAN WHITE PELICAN.

An abundant summer resident of the Western Plains, and north to about the 60th parallel.

It is very rare in British Columbia, occurring only between the Cascade Mountains and the Rockies. Small flocks are seen occasionally in the Great Lakes, and one specimen has been taken in Nova Scotia and two in New Brunswick. It winters in southern latitudes.

Pelecanus californicus.
CALIFORNIA BROWN PELICAN.

A few examples have been observed along the coast of British Columbia.

Fregata aquila.
MAN-O'-WAR BIRD.

Two of these birds have been taken in Canadian waters — one off Halifax, October 16, 1876, and one off Point des Monts, P. Q., by Mr. COMEAU, August 13, 1884.

Merganser americanus.
AMERICAN MERGANSER.

A common species throughout the Dominion, breeding from about the 45th parallel to the Hudson's Bay region. Some winter in New Brunswick and in Southern Ontario.

Merganser serrator.
RED-BREASTED MERGANSER.

This is the "Shell-drake" of sportsmen, and is distributed over the entire Dominion, though more abundant near the Atlantic and Pacific coasts than in the interior. It breeds from about the 45th parallel to the Arctics, but none winter in Canada.

Lophodytes cucullatus.
HOODED MERGANSER.

An abundant summer resident of Canada, breeding throughout its range (which extends to the Fur Countries), excepting in Southern Ontario, where it occurs as a migrant only.

Anas boschas.
MALLARD.

This species breeds abundantly in British Columbia, and is abundant eastward to the longitude of Hamilton, from which point its numbers decrease, until—in the Maritime Provinces—very few examples are met with. It goes as far north as Hudson's Bay, and south, in winter, to Central America.

Anas obscura.
BLACK DUCK.

Abundant in the eastern half of Canada—especially in the Maritime Provinces—but rare in Manitoba, and absent from the Plains.

It breeds as far north as Central Labrador, and south to the St. Clair Flats, Ontario.

Anas strepera.
GADWALL.

Mr. THOMPSON has reported this species as rare in Manitoba, though Dr. COUES, when on the boundary survey, found it abundant along the Red River Valley, and thence to the Rockies. Prof. MACOUN reports it as abundant in the Lakes of Assiniboia and Saskatchewan, and Mr. FANNIN considers it fairly abundant in British Columbia; while from Manitoba to the Atlantic it is very seldom met with.

It has been taken in Slave Lake and Hudson's Bay.

Anas penelope.
WIDGEON.

A straggler from the Eastern Hemisphere, occasionally found along the Atlantic coast.

Anas americana.
BALDPATE.

This is the species that is usually called "Widgeon" by sportsmen. It is abundant on the Pacific slope, and eastward to the Ottawa Valley, from which point to the Atlantic it is rather uncommon, and is considered a rare bird by many observers in New Brunswick and Nova Scotia.

Anas crecca.
EUROPEAN TEAL.

One example of this species was taken near Halifax, by Dr. GILPIN, on September 1, 1854, and Dr. COUES took one in Labrador.

Anas carolinensis.
GREEN-WINGED TEAL.

Common throughout most of the Dominion, though uncommon in the Maritime Provinces, and abundant in British Columbia. It breeds from about latitude 50° to the Fur Countries, and winters in Central America.

Anas discors.
BLUE-WINGED TEAL.

An abundant species along the southern portions of the Dominion from the Atlantic to the Pacific, ranging as far north as Great Slave Lake in the west; but along the Atlantic sea-board is rarely seen north of the St. Lawrence. It winters in Central America.

Anas cyanoptera.
CINNAMON TEAL.

A rare summer resident of British Columbia; reported by THOMPSON, on the authority of Mr. R. H. HUNTER, as occurring occasionally in Manitoba. Winters in South America.

Spatula clypeata.
SHOVELLER.

This species is very abundant on the Plains, from Manitoba to the Rockies, but on the Pacific slope it is rare; on the Atlantic border it is common in a few localities, but generally rather uncommon. It breeds from our southern boundary to the Fur Countries, and winters in the Southern States.

Dafila acuta.
PINTAIL.

A rather rare bird along both ocean borders, but common in the interior, and abundant on the Plains; breeds from about the 45th parallel to the Lower Fur Countries. Winters in Cuba and Central America.

Aix sponsa.
WOOD DUCK.

Occurs throughout the Dominion, though rare on the Plains and north of latitude 50°; probably not found beyond the 60th parallel.

Netta rufina.
RUFOUS-CRESTED DUCK.

A bird of the Eastern Hemisphere that occurs accidentally along the Atlantic coast.

Aythya americana.
REDHEAD.

Occurs throughout the Dominion, being least common in the Maritime Provinces, and abundant from Montreal to Western Manitoba. On the Atlantic coast it is not found north of the St. Lawrence, but in the west it ranges as far north as the Fur Countries. A few breed in Southern Ontario.

Aythya vallisneria.
CANVAS-BACK.

Occurs throughout the Dominion, but is merely occasional in the Maritime Provinces, while a common bird in Lake Manitoba and the western portion of the Great Lakes. In the west it ranges to Great Slave Lake and Alaska.

Aythya marila nearctica.
AMERICAN SCAUP DUCK.

This species is fairly common throughout the Dominion, breeding from about latitude 50° northward, though a few pairs are said to nest on the St. Clair Flats, Ontario, in latitude 43°.

Aythya affinis.
LESSER SCAUP DUCK.

This species, which is better known to Canadian gunners as the "Little Black-head," occurs throughout the Dominion, and is abundant everywhere excepting in the Maritime Provinces, where it is only "uncommon." It is not reported from Labrador, but Dr. BELL found it "breeding in large numbers" on Nottingham Island, in Hudson's Bay. It winters in the West Indies.

Aythya collaris.
RING-NECKED DUCK.

According to Mr. THOMPSON's list of Manitoba birds, this species is abundant in the Red River Valley, and Mr. FANNIN reports it abundant in British Columbia, but elsewhere in Canada it appears to be of rather rare and irregular occurrence. It has been taken as far north as Fort Simpson, and winters in the West Indies.

Glaucionetta clangula americana.
AMERICAN GOLDEN-EYE.

The "Whistler" of Canadian sportsmen. It is common everywhere throughout the Dominion, breeding, generally, in the far north. Mr. Joseph Macoun observed it passing Lake Misstassini on May 3. It winters south to Cuba.

Glaucionetta islandica.
BARROW'S GOLDEN-EYE.

This species usually breeds in the far north, though Dr. Coues found a brood of young, in August, in the Rocky Mountains, near the 49th parallel, and it is said to remain during the summer near Point des Monts, in the Gulf of St. Lawrence. A nest with eggs was taken by Mr. C. C. Beattie at Missisquoi, Lake Champlain; and the male, which was shot near the nest, was identified by Mr. William Couper, of Montreal.

The species is rarely seen in the Great Lakes, and is uncommon in British Columbia and Manitoba, as well as in the Atlantic Provinces. In the latter it occurs as a winter visitor only.

Charitonetta albeola.
BUFFLE-HEAD.

A common bird throughout Canada — especially in the west, where it is sometimes very abundant. It breeds over its entire range in this country, though somewhat sparingly in the more southern portions.

Clangula hyemalis.
OLD-SQUAW.

An abundant bird from the Atlantic to the Pacific; breeding in the far north, and wintering in the Southern States.

Histrionicus histrionicus.
HARLEQUIN DUCK.

According to Mr. FANNIN this bird is abundant in British Columbia, and Dr. COUES reports it breeding in the Rocky Mountains, near the United States boundary. Between the Rockies and Montreal it occurs accidentally, the only records being of three examples taken by Mr. LOANE, near Toronto (*Thompson*), and a few observed near Montreal. On the Atlantic coast it is not common, and occurs usually as a migrant, a few wintering in the Bay of Fundy.

Camptolaimus labradorius.
LABRADOR DUCK.

Mr. BOARDMAN reports that this species was quite common in the Bay of Fundy in 1845, but it has since become very rare, and, possibly, may be extinct. A female taken at Grand Manan in 1871 is the last recorded capture, though one is said to have been taken off Nova Scotia a few years later.

Eniconetta stelleri.
STELLER'S DUCK.

KUMLIEN reports having seen a few of this species in Cumberland bay, and of shooting one at Disko.

Somateria mollissima.
EIDER.

An abundant bird along the Atlantic sea-board, from Hudson's Strait northward, and occasionally found, in winter, at the mouth of the Bay of Fundy.

Somateria dresseri.

AMERICAN EIDER.

A common bird along our sea-board, from the Bay of Fundy to Labrador, where it breeds. It is generally reported as rare in the Great Lakes, though the Rev. VINCENT CLEMENTI considers it fairly common near Peterborough, Ontario.

Somateria v-nigra.

PACIFIC EIDER.

This species is found along the northern coast of British Columbia, and in Dease Lake and Great Slave Lake.

Somateria spectabilis.

KING EIDER.

According to Mr. BOARDMAN this species formerly bred on Grand Manan, and Mr. COMEAU and Mr. COUPER have found it breeding along the north shore of the Gulf of St. Lawrence; but it usually goes farther north to nest, occurring, as a winter visitor only, south of Labrador. During the winter it also occurs, occasionally, in the Great Lakes; several flocks were seen near Buffalo and Niagara in 1879.

Oidemia americana.

AMERICAN SCOTER.

A common bird everywhere, excepting on the Western Plains; going to the far north to breed, and wintering from the United States border southward.

Oidemia deglandi.
WHITE-WINGED SCOTER.

The "White-winged Coot" of our gunners. A common bird throughout the entire Dominion, breeding from about latitude 50° northward to the Arctics. Winters in the Middle States and Southern California.

Oidemia perspicillata.
SURF SCOTER.

This species is common during the migrations on both sea coasts, but is rarely seen in the interior, though it breeds in Great Slave Lake and in Hudson's Bay, where Dr. BELL found it in "immense numbers." On the Atlantic coast it breeds from the north shore of the Gulf of St. Lawrence to Northern Labrador. It winters in the Southern States and Lower California.

Erismatura rubida.
RUDDY DUCK.

This species is more abundant in the interior than on either sea-board. It occurs only as a migrant in the Maritime Provinces, but spends the summer in Manitoba, and Dr. COUES found it nesting at Turtle Mountain, near latitude 49°. In winter it goes to Cuba and northern South America.

Chen hyperborea.
LESSER SNOW GOOSE.

An abundant bird on the mainland of British Columbia, but uncommon on Vancouver Island. It has been found in Illinois; Dr. BELL records one taken at Fort Churchill; and Mr. BOARDMAN has a specimen taken in the Bay of Fundy. It breeds in Alaska, and winters in Southern California.

Chen hyperborea nivalis.
GREATER SNOW GOOSE.

This species is more abundant on the Plains than nearer the Atlantic coast, comparatively few having been found east of Manitoba, while on the sea-board it is extremely rare. It breeds far north, and winters along the coast of the South Atlantic States.

Chen rossii.
ROSS'S SNOW GOOSE.

A rather common species in British Columbia, though more abundant in the interior than on the coast. It has been taken in Great Slave Lake and at Fort Churchill, and winters in Southern California.

Anser albifrons gambeli.
AMERICAN WHITE-FRONTED GOOSE.

This species appears to be met with more frequently in British Columbia and in the vicinity of the Great Lakes than elsewhere in the Dominion. One pair taken at Turtle Mountain, in 1882, is the only record for that district, and DUNLAP and WINTLE's report of its accidental occurrence near Montreal is the only eastern record, excepting KUMLIEN's report of finding it in Cumberland Bay and in Greenland. It winters in Mexico and Cuba.

Branta canadensis.
CANADA GOOSE.

The "Wild Goose" occurs throughout the entire Dominion, but is more abundant between the Rockies and the Atlantic than on the Pacific slope. It breeds from about the 50th parallel northward, and in winter goes south to Mexico.

Branta canadensis hutchinsii.
HUTCHINS'S GOOSE.

A north-western form. It breeds in the Arctic regions, and goes south in winter. It is rather common in British Columbia, and occurs casually in the Great Lakes, and in Hudson's Bay, and was taken by KENNICOTT in Red River. KUMLIEN reports taking one in Cumberland Bay.

Branta canadensis occidentalis.
WHITE-CHEEKED GOOSE.

Occurs along the sea-coast of British Columbia, but is not a common bird. It winters in California.

Branta canadensis minima.
CACKLING GOOSE.

This species must pass through Canadian territory on its migrations from Alaska to the Western States, but there is, as yet, no record of its capture within the Canadian boundaries.

Branta bernicla.
BRANT.

A very abundant migrant along the Atlantic coast, but neither COUPER nor NEILSON have seen it near Quebec, and it is but rarely met with at Montreal, or through the Great Lakes, though GIBBS gives it as "common" in Michigan. Its name appears in THOMPSON's catalogue of Western Manitoba, and in Dr. COUES's list of species observed in Northern Dakota and Montana. It probably breeds in the interior of Labrador and in the large lakes near Cumberland Bay.

Branta nigricans.
BLACK BRANT.

Very abundant along the coast of British Columbia during the migrations.

Branta leucopsis.
BARNACLE GOOSE.

A few examples have been taken in the Hudson's Bay district, but it is an accidental straggler there from the Eastern Hemisphere.

Olor columbianus.
WHISTLING SWAN.

This species is more or less common from the Pacific coast to Western Ontario, but from the St. Clair Flats eastward it is only casual. LUCIEN TURNER gives it as occasional in Labrador; FRANCIS BAIN reports one taken in Prince Edward Island in 1885; and one was taken in New Brunswick in 1882. The Indians of the latter Province are familiar with the appearance of the Swan—having a name for it—and say they find it occasionally in the wake of flocks of Geese.

Olor buccinator.
TRUMPETER SWAN.

This Swan is rather common in British Columbia, but eastward of that Province it is rare or casual.

Mr. THOMPSON reports that there is no record of its occurrence in Manitoba, but Dr. COUES gives it as occasional in the Red River Valley. Rev. Mr. HINCKS, of Toronto, calls it "our commonest species"; Mr. STOCKWELL, in *Forest and Stream*, reports it as "occasional" in the St. Clair Flats, and Mr. McILWRAITH states that a few of this species have been taken in Ontario. Dr. WHEATON gives it as "a rare winter visitor" to Ohio. Dr. ROBERT BELL reports its occurrence at several points in Hudson's Bay, and says it breeds on the Islands in the Bay.

It breeds from the Yellow-Stone Park to the Fur Countries.

Plegadis autumnalis.
GLOSSY IBIS.

An accidental straggler. Dr. BREWER mentions that Mr. FRANK L. TILESTON saw several birds, "undoubtedly of this species," on Prince Edward Island in 1878; Mr. MCILWRAITH reports that Mr. JOHN BATES took two near Hamilton in 1857; and Mr. FANNIN has taken one at Salt Springs Island, off British Columbia.

Botaurus lentiginosus.
AMERICAN BITTERN.

A common summer resident throughout Canada to about the 60th parallel.

Botaurus exilis.
LEAST BITTERN.

This is said to be a common bird on the Western Plains, though THOMPSON reports that only one example is known to have been seen in Manitoba. It is quite common in Southern Ontario, but east of that locality is merely accidental. A few have been taken along the New Brunswick shore of the Bay of Fundy.

Ardea herodias.
GREAT BLUE HERON.

A common summer resident of the southern portions of Canada, being especially abundant in British Columbia. Near the Atlantic it is seldom seen north of latitude 48°, though common near latitude 45°.

Ardea egretta.

AMERICAN EGRET.

Mr. SAUNDERS considers that this is a regular, though rare, visitor to Southern Ontario; elsewhere in Canada it is merely accidental. Mr. SCOTT reports that two specimens were shot in the Upper Ottawa, in latitude 47° 50'; Mr. HERRCHFELDER secured another in Lake Nippissing; Mr. COMEAU has seen one at Point des Monts; Mr. JONES records one taken at Halifax, and three have been captured in New Brunswick.

Ardea candidissima.

SNOWY HERON.

An accidental straggler from the south. Mr. JONES reports that one specimen has been taken at Windsor, N. S.; Mr. BOARDMAN has two that were captured near the mouth of the Bay of Fundy; Mr. FANNIN has taken one at Burrard Inlet, and another in the Fraser River, and Mr. MCILWRAITH, on the authority of Dr. GARNIER, reports the capture of one at Mitchell's Bay.

This last note is inserted in *The Birds of Ontario* under the heading of *A. egretta*, but, as Dr. GARNIER is quoted as naming the specimen "*Garzetta candidissima*—Little White Heron," it is obvious that it should be referred to the present species. Mr. MCILWRAITH reports that the error was made by the printer.

Ardea cœrulea.

LITTLE BLUE HERON.

Mr. J. MATTHEW JONES reports that one example of this species was taken at Cole Harbour, near Halifax, during the summer of 1884.

Ardea virescens.
GREEN HERON.

This species occurs regularly in the Eastern Provinces, though it is usually rather rare, but occasionally it has been fairly common in Southern Ontairo, and on the shores of the Bay of Fundy. Prof. MACOUN reports finding it in the Assiniboine River.

Nycticorax nycticorax nævius.
BLACK-CROWNED NIGHT HERON.

This species is rather common in many localities throughout the southern portions of Canada, between the Atlantic and Western Manitoba. On the Atlantic coast it has not been taken north of Gaspé, though Mr. JOSEPH MACOUN found one specimen at Lake Misstassini. Mr. SCRIVEN reports it common in the Muskoka District.

Grus americana.
WHOOPING CRANE.

A common bird on the Western Plains. Mr. WILLIAM LOANE reports having seen one near Toronto (*Thompson*).

Grus canadensis.
LITTLE BROWN CRANE.

A western species, breeding in the Fur Countries, and wintering in the Western United States.

Grus mexicana.
SANDHILL CRANE.

This species is abundant on the Plains, and is common in British Columbia. A few examples have been taken in Ontario, and in the Hudson's Bay district.

E

Rallus elegans.
KING RAIL.

Both in the SAUNDERS–MORDEN list and in Mr. MCILWRAITH'S *Birds of Ontario* this species is given as common on the St. Clair Flats; Mr. CLEMENTI reports it fairly common near Peterborough; and in the DUNLAP-WINTLE list of Montreal species it is given as "tolerably common."

I have examined in the flesh one specimen that was shot near St. John, N. B., in June, 1887.

Rallus virginianus.
VIRGINIA RAIL.

Occurs throughout the Dominion. It is rare in British Columbia, and seldom seen north of latitude 50°, though Dr. BELL reports its occurrence at York Factory, and TURNER took one at Hamilton Inlet, Labrador.

Porzana carolina.
SORA.

The "Carolina Rail" is well known to all sportsmen between the Atlantic and Western Manitoba. It ranges as far north as the Hudson's Bay district, and south to northern South America.

Porzana noveboracensis.
YELLOW RAIL.

This species occurs occasionally throughout the Eastern Provinces, but is not common anywhere. It has been taken in the Hudson's Bay district, and THOMPSON reports having examined one specimen taken in Manitoba.

THE CORN CRAKE (*crex crex*), a European species, is said to occur in Eastern North America, but I have seen no record of its having been observed in Canadian territory.

Ionornis martinica.
PURPLE GALLINULE.
The claim of this species to be named in the Canadian fauna rests on the capture of one example near Halifax, N. S., and another near St. John, N. B.

Gallinula galeata.
FLORIDA GALLINULE.
This species is common in Southern Ontario, and occurs regularly at Ottawa (*Scott*) and Montreal (*Dunlap* and *Wintle*), but it is merely accidental in the Maritime Provinces.

Fulica americana.
AMERICAN COOT.
A common bird throughout the Dominion to about the 50th parallel, going farther north in the west than on the Atlantic seaboard.

Crymophilus fulicarius.
RED PHALAROPE.
A common bird along the Atlantic, breeding north of Labrador, and migrating south in winter. Mr. BOARDMAN thinks a few breed in the Bay of Fundy. It has been occasionally seen in the Great Lakes; Dr. GARNIER reports shooting one from a flock of six at Mitchell's Bay, and Mr. BROOKS captured one near Hamilton.

Phalaropus lobatus.
NORTHERN PHALAROPE.
This species is common along both sea coasts. On the Atlantic it has not been found breeding south of Ungava Bay,

but Mr. FANNIN reports its occurrence during the breeding season at Burrard Inlet, near the southern border of British Columbia. It occurs sparingly on the Plains during the migrations, and also appears occasionally in the Great Lakes.

Phalaropus tricolor.
WILSON'S PHALAROPE.

A common bird on the Western Plains, breeding along our southern border—from Manitoba to the Rockies, and north to the Saskatchewan. WHEATON gives it as a not uncommon spring and fall migrant through Ohio, but the observers in Ontario consider it merely accidental in that Province.

Recurvirostra americana.
AMERICAN AVOCET.

This species is very abundant in all the saline regions of the Western Plains, and has been taken as far north as Great Slave Lake. MCILWRAITH mentions that three examples have been taken in Ontario, and a few have been taken in New Brunswick.

Himantopus mexicanus.
BLACK-NECKED STILT.

Dr. WHEATON reports that this species has been repeatedly taken in Lake Erie, but it is not given in any Ontario list, though THOMPSON reports that he has heard of it being taken near Toronto. Dr. GIBBS gives it as rare in Michigan. A few specimens have been taken in New Brunswick. It is a southern and western bird, going, in winter, as far south as Brazil and Peru.

Scolopax rusticola.
EUROPEAN WOODCOCK.

An example of this species was taken at Chambly, P. Q., on March 11, 1882. It was identified by Mr. BROCK WILLET and Mr. COUPER.

Philohela minor.
AMERICAN WOODCOCK.

Common in the Maritime Provinces and west to Lake Huron. THOMPSON gives it as rare in Manitoba, though Dr. BELL reports finding it not uncommon along the Red River Valley. It does not range far northward; SCRIVEN reports it rare at Gravenhurst, and Dr. BELL's report of finding one specimen at York Factory, and Mr. WILLIAM BREWSTER's report of one at Gaspé, are the most northern records, though TURNER heard that several had been killed in Eastern Labrador.

THE EUROPEAN SNIPE (*gallinago gallinago*), has been taken in Greenland and in Bermuda, but there is no record of its occurrence within the boundaries of Canada.

Gallinago delicata.
WILSON'S SNIPE.

An abundant bird from the Atlantic to the Pacific, though of rather irregular distribution. It breeds from our southern border northward to Labrador and the Hudson's Bay district.

Macrorhamphus griseus.
DOWITCHER.

This species is reported by MACOUN as abundant on the Western Plains, though THOMPSON gives it as a rare migrant through Manitoba. (Dr. Coues concludes, from his observations, that some may breed in the vicinity of latitude 49°). It is common on the Atlantic coast during the autumn, and rare along the Great Lakes in spring and autumn.

Macrorhamphus scolopaceus.
LONG-BILLED DOWITCHER.

An abundant bird in British Columbia, and said to occur on the Plains and along the Atlantic coast, but I can find no record of any being taken in Canada east of the Rockies.

Micropalama himantopus.
STILT SANDPIPER.

This is a rare bird everywhere, and the records of its occurrence in Canada are but few.

Dr. Coues took some half-dozen specimens near the Rockies in the vicinity of the 49th parallel; Prof. Macoun secured one near the Qu'Appelle River; a few have been taken in the Muskoka district (*Scriven*); two near Toronto (*Thompson*); and three were obtained by Mr. F. W. Daniel near St. John, N. B. Dr. Gibbs reports that two were taken in Lake Michigan, and Dr. Wheaton tells of a few being taken in Lake Erie.

It is said to range north to the Fur Countries.

Tringa canutus.
KNOT.

This bird is sometimes called "Robin Snipe" by sportsmen. It occurs throughout Canada, being most numerous on the Prairies and in British Columbia; breeding in the Arctic regions to latitude 81°. It winters in South America.

Tringa maritima.
PURPLE SANDPIPER.

An abundant species along the Atlantic and Pacific coasts, but rare in the interior. The only records for Ontario are five specimens taken near Hamilton (*McIlwraith*), one at Toronto (*Thompson*), and one at Ottawa (*White*). It is accidental, also, along the St. Lawrence to Quebec, and has not been met with by any observer on the Great Plains. In the Maritime Provinces it is a winter resident.

Tringa maculata.
PECTORAL SANDPIPER.

This species is commonly called "Jack Snipe" by our gunners. It occurs throughout Canada, migrating to its breeding ground in the Arctics by the Mississippi Valley route and along the Pacific coast, and in autumn appearing in large flocks along the Atlantic. In the Great Lakes it is common in spring and autumn. It winters in South America.

Tringa fuscicollis.
WHITE-RUMPED SANDPIPER.

Common in Manitoba in the spring, and common along the Atlantic coast in autumn. It is only occasionally seen in the Great Lakes. Breeds in high latitudes, and winters in South America.

Tringa bairdii.
BAIRD'S SANDPIPER.

An abundant migrant on the Plains, from Manitoba to the Rockies. Dr. WHEATON reports that several have been taken in Lake Erie in spring and fall, and Mr. McILWRAITH mentions two examples in his *Birds of Ontario.*

Tringa minutilla.

LEAST SANDPIPER.

Occurs throughout Canada, but is most abundant along the Atlantic border, where it appears in autumn in large flocks, and where it is occasionally seen in spring. It breeds from Labrador to the Arctic Ocean, and winters in South America.

Tringa alpina.

DUNLIN.

An accidental straggler from the old world. It has been taken in the Hudson's Bay district by Captain BLAKISTON, and British Columbia by Mr. ELLIOTT and Mr. BROWN.

Tringa alpina pacifica.

RED-BACKED SANDPIPER.

This is the western form of the Dunlin. It is usually rare on the Atlantic border, though common between Montreal and the Western Plains, and abundant along the Pacific coast. It breeds in the far north.

Tringa ferruginea.

CURLEW SANDPIPER.

The usual habitat of this species is in the Eastern Hemisphere, though KUMLIEN found it not uncommon in Northern Greenland. BOARDMAN has three specimens taken in the Bay of Fundy; DOWNS has one which he procured in the Halifax market, and one is in the rooms of the GUN CLUB, Toronto, that was taken in Southern Ontario.

Ereunetes pusillus.
SEMIPALMATED SANDPIPER.

A common spring and autumn migrant between Lake Ontario and the Rockies, and very abundant on the borders of the Atlantic in autumn. It breeds in the far north, and winters in the West Indies and South America.

Ereunetes occidentalis.
WESTERN SANDPIPER.

A fairly common bird in British Columbia, from the Rockies to the Pacific, breeding in high latitudes.

Calidris arenaria.
SANDERLING.

A common bird throughout Canada, breeding from about latitude 55° to latitude 82°. Goes south in winter to Chili and Patagonia.

Limosa fedoa.
MARBLED GODWIT.

Occurs from the Atlantic to the Pacific, but most abundant on the Western Plains, and rare along the Atlantic border. Breeds from about latitude 50° northward, and winters in Central America.

Limosa lapponica baueri.
PACIFIC GODWIT.

Mr. FANNIN reports that this is a common winter visitor along the lower Fraser River, B. C.

Limosa hæmastica.
HUDSONIAN GODWIT.

Occurs from the Atlantic to the Rockies, and north to the Fur Countries; most common along the Atlantic border, and rarest on the Plains.

Totanus melanoleucus.
GREATER YELLOW-LEGS.

Occurs throughout the entire Dominion to sub-Arctic regions. AUDUBON found it breeding in Labrador, and BREWSTER thinks it breeds on Anticosti, where he found it very abundant. Mr. HIND thinks it breeds in the northern lakes of Manitoba. It winters south to Chili and Buenos Ayres.

Totanus flavipes.
YELLOW-LEGS.

Occurs throughout the entire Dominion to sub-Arctic regions, though rare on the Atlantic coast in the spring. Winters from West Indies to southern South America.

Totanus solitarius.
SOLITARY SANDPIPER.

Occurs as a summer resident throughout Canada, north to the Fur Countries; south in winter to the tropics.

Totanus ochropus.
GREEN SANDPIPER.

An accidental straggler from the Old World that has been taken in Nova Scotia.

Symphemia semipalmata.

WILLET.

A common summer resident of the plains west of Manitoba, but rare in the latter Province; a rare migrant in the vicinity of the Great Lakes; and, usually, a common autumn visitor to the Maritime Provinces. Dr. BRYANT received a set of eggs from Mr. DOWNS that were said to have been taken near Yarmouth, N. S.

p. 123.

Heteractitis incanus.

WANDERING TATLER.

Mr. FANNIN has taken this species both east and west of the Cascade Mountains in British Columbia.

Pavoncella pugnax.

RUFF.

An accidental straggler from the Old World. One or two specimens have been taken on the Bay of Fundy shores, and one was captured near Toronto.

Bartramia longicauda.

BARTRAMIAN SANDPIPER.

This species was formerly called the "Field Plover." Along the Atlantic border it occurs north to the Gulf of St. Lawrence, and on the Pacific it ranges to Alaska. It is most abundant in the Prairie region, and rarest along the Great Lakes. It winters in the tropics.

Tryngites subruficollis.
BUFF-BREASTED SANDPIPER.

Mr. FANNIN reports that this species is abundant, in winter, at the mouth of the Fraser River, B. C., and Mr. THOMPSON gives it as a very rare fall migrant in Manitoba. The more eastern records are: several seen near Hamilton in autumn (*McIlwraith*); a pair breeding near Dunnville, Ont., in 1879 (*McCallum*); a male captured near Ottawa, August 24, 1886 (*G. R. White*); rare near Montreal (*Hall*); several taken near Montreal, in May, 1884 and 1885 (*Keutzing*); one taken by COMEAU near Point des Monts (*Merriam*); one taken on Prince Edward Island (*Tileston*); one taken at Port Burwell, Hudson's Bay (*Bell*), and one taken in Labrador (*Coues*). It has been found breeding along the Anderson River, and the Yukon; and winters in South America.

Actitis macularia.
SPOTTED SANDPIPER.

A common bird from the Atlantic to the Pacific, breeding along our southern border and northward. TURNER found it at Fort Chimo, Labrador, and Dr. BELL reports it occurring in the Hudson's Bay district.

Numenius longirostris.
LONG-BILLED CURLEW.

This species occurs in the eastern portions of Canada as an uncommon autumn migrant, but it is fairly common in British Columbia, where it is found during the entire summer. Dr. COUES reports it breeding near Pembina in moderate numbers. MCILWRAITH gives it as an occasional migrant in Ontario, though THOMPSON reports that, while it is rare near Toronto in the autumn, it passes that locality during the spring migrations in very large flocks, and SAUNDERS is of the opinion that it occurs in Ontario regularly and in considerable numbers. There is no record of its occurrence on the Atlantic coast north of the Baie de Chaleur, where it appears in autumn only.

Numenius hudsonicus.
HUDSONIAN CURLEW.

An abundant bird, in the migrations, on the Atlantic coast up to Anticosti, but rare on the north side of the St. Lawrence river. It is a rare migrant along the Great Lakes, and occurs occasionally at Ottawa and Montreal. Dr. BELL found it abundant at Fort Churchill, and it has been reported from the Slave Lake region. It winters far south.

Numenius borealis.
ESKIMO CURLEW.

This species is common along the Atlantic coast during the migrations, and is fairly common some seasons at Quebec and Montreal. McILWRAITH reports that only one has been taken in Ontario; THOMPSON maintains that two have been seen near Toronto, and Dr. WHEATON has put it down as a regular migrant along Lake Erie. Dr. BELL reports finding it abundant at Fort Churchill in July and August, and KUMLIEN, when at Cumberland Bay, saw several flocks going north in June. It migrates in winter to the southern extremity of South America.

Charadrius squatarola.
BLACK-BELLIED PLOVER.

A common species throughout the entire Dominion, occurring chiefly as a spring migrant on the Prairies, and as an autumn migrant along the Atlantic coast. It breeds in the far north, and winters in the West Indies and South America.

Charadrius dominicus.
AMERICAN GOLDEN PLOVER.

Occurs throughout the entire Dominion, but is more abundant on the Plains than along the ocean borders. Some observers in Ontario think it rare or absent, but others think it common, and this latter opinion is confirmed by Dr. WHEATON's experience in Ohio. It ranges from the Arctics to Patagonia.

Ægialitis vocifera.
KILLDEER.

This species is rather rare on both sea coasts, but is common along the Great Lakes, and abundant on the Prairies. It breeds in the temperate regions, and winters in South America.

Ægialitis semipalmata.
SEMIPALMATED PLOVER.

Occurs throughout the Dominion, but is most abundant along the Atlantic coast. It breeds in the Fur Countries and northward, and winters in the tropics.

Ægialitis hiaticula.
RING PLOVER.

A bird of the Old World that KUMLIEN found breeding in Cumberland Bay.

Ægialitis dubia.
LITTLE RING PLOVER.

Mr. FANNIN reports taking this species on Vancouver Island.

Ægialitis meloda.
PIPING PLOVER.

This species is reported as breeding abundantly on the Magdalene Islands by Mr. CORY, and Prof. McKAY has found it breeding near Pictou, N. S.; it is said to breed, also, on the islands in the Bay of Fundy, though it is seen in that vicinity but occasionally. McILWRAITH has met with it in Ontario on two occasions only, though THOMPSON considers it a common migrant near Toronto, and SAUNDERS found it breeding at Point Pelee. It has been seen along the shores of Lake Manitoba and Lake Winnipeg. Winters in the West Indies.

Ægialitis nivosa.
SNOWY PLOVER.

Mr. FANNIN has taken this species at the mouth of the Fraser River in winter; and Mr. THOMPSON reports that one specimen was shot by Mr. I. FORMAN, near Toronto, in May, 1880.

Ægialitis wilsonia.
WILSON'S PLOVER.

Col. Goss reports shooting one of these birds at Brier Island, N. S., on April 28, 1880. Mr. THOMPSON reports it a very rare bird near Toronto, but Mr. MORDEN gives it as a regular visitor to the vicinity of Hyde Park, Ontario.

Ægialitis montana.
MOUNTAIN PLOVER.

Dr. Coues reports finding this species abundant between the Sweet-grass Hills and the Milk River, in latitude 49°.

Aphriza virgata.
SURF BIRD.

This species occurs along the whole coast line of British Columbia, but it is most abundant north of Vancouver Island.

Arenaria interpres.
TURNSTONE.

Occurs throughout Canada, though generally reported somewhat rare in the interior, excepting near Toronto, where Mr. THOMPSON has found it in immense numbers in the spring, though less abundant in the autumn. It breeds in high latitudes, and winters far south.

Arenaria melanocephala.
BLACK TURNSTONE.

This species occurs in British Columbia, from Burrard Inlet northward, but is most abundant in Howe Sound (*Fannin*).

Hæmatopus palliatus.
AMERICAN OYSTER-CATCHER.

An accidental straggler from the south. Two specimens have been taken at the mouth of the Bay of Fundy: one on Grand Manan, and another on the mainland, near Eastport.

Hæmatopus bachmani.
BLACK OYSTER-CATCHER.
This species occurs on the coast of British Columbia, from Cape Flattery to the Alaskan boundary, being most abundant at Howe Sound. It has not been obseved at Burrard Inlet (*Fannin*).

Colinus virginianus.
BOB-WHITE.
Occurs in Ontario only. MCILWRAITH calls it a common bird, and SAUNDERS and MORDEN hold the same opinion, though it is reported as rare from Toronto (*Thompson*), Lucknow (*Garnier*), and Gravenhurst (*Scriven*).

Oreortyx pictus.
MOUNTAIN PARTRIDGE.
This species has been introduced on Vancouver Island.

Callipepla californica vallicola.
VALLEY PARTRIDGE.
This species has been introduced on Vancouver Island.

Dendragapus obscurus fuliginosus.
SOOTY GROUSE.
Occurs in British Columbia, between the Cascade Mountains and the Pacific, and on the Islands along the coast.

Dendragapus obscurus richardsonii.
RICHARDSON'S GROUSE.
Occurs in British Columbia, on the eastern slope of the Cascades, and in the Rockies. Occurs, also, along the Saskatchewan,

G

and Mr. C. J. BAMPTON reports having frequently seen specimens in the market at Sault St. Marie (*McIlwraith*).

Dendragapus canadensis.
CANADA GROUSE.

This is the "Spruce Partridge" of eastern sportsmen. It occurs from the Atlantic to the Rockies, and from the 45th parallel to the Fur Countries.

Dendragapus franklinii.
FRANKLIN'S GROUSE.

An abundant species in the timbered districts of British Columbia east of the Cascades, from the southern boundary to the Arctic water-shed. Dr. COUES found it on the eastern slope of the Rockies.

Bonasa umbellus togata.
CANADIAN RUFFED GROUSE.

This is the "Birch Partridge" of eastern sportsmen. It occurs in the timbered districts of the eastern Provinces, and west to the Rockies, and from the southern boundary to the Fur Countries. (Mr. THOMPSON considers that the bird found in Manitoba should be referred to *umbelloides*).

Bonasa umbellus umbelloides.
GRAY RUFFED GROUSE.

This form of the "Birch Partridge" occurs in the interior of British Columbia, from the southern boundary to Alaska (probably, also, in Manitoba).

Bonasa umbellus sabini.
OREGON RUFFED GROUSE.

This is the form of the "Birch Partridge" that is found on the Pacific coast. It occurs in British Columbia, from the summit of the Cascades to the sea-shore, and on the islands along the coast.

Lagopus lagopus.
WILLOW PTARMIGAN.

This species occurs from the Atlantic to Lake Winnipeg, and north to the Arctics. It is very abundant at Point des Monts, P. Q., and occurs at Sault St. Marie; but the only specimens that are known to have been taken south of the St. Lawrence are one captured in the Magdalene Islands (*Cory*), and another in Lewis County, N. Y. (*Merriam*). In Dr. HALL's list of Montreal birds it is given as rare, but later observers have failed to procure any evidence of its occurrence in that vicinity, though Mr. NIELSON has found it near Quebec, and Dr. BRODIE reports it from thirty miles north of Toronto. It is probably most abundant in Labrador and in the Hudson's Bay district.

ALLEN'S PTARMIGAN (*Lagopus lagopus alleni*) is a form of the Willow Ptarmigan that occurs in Newfoundland only.

Lagopus rupestris.
ROCK PTARMIGAN.

TURNER reports this as abundant on the treeless area of Labrador, where it remains the entire year, and Dr. BELL found it in vast numbers around Hudson's Strait. It has been taken near Quebec (*Couper*); at Sault St. Marie (*McIlwraith*); and on Anticosti (*Brewster*).

Lagopus rupestris reinhardti.
REINHARDT'S PTARMIGAN.

This species occurs on the western shores of Cumberland Bay.

Lagopus leucurus.
WHITE-TAILED PTARMIGAN.

An abundant resident of the eastern slopes of the Cascades. It occurs north to Dease Lake, and south to New Mexico.

Tympanuchus americanus.
PRAIRIE HEN.

This species has lately become common on the Plains of Manitoba, and it occurs, also, in Southern Ontario.

Pediocætes phasianellus.
SHARP-TAILED GROUSE.

In the A. O. U. check list the habitat of this species is given as " British America, from the northern shores of Lake Superior, and British Columbia, to Hudson's Bay Territory and Alaska."

Pediocætes phasianellus columbianus.
COLUMBIAN SHARP-TAILED GROUSE.

Mr. FANNIN reports that this form occurs in British Columbia, between the Cascades and the Rockies. Mr. THOMPSON reports it very abundant in Manitoba.

Pediocætes phasianellus campestris.
PRAIRIE SHARP-TAILED GROUSE.

This sub-species is said to occur on the Great Plains.

Centrocercus urophasianus.
SAGE GROUSE.

Mr. FANNIN reports that this species occurs occasionally in the open hills along the southern border of British Columbia, and Dr. COUES found it on the Plains between the Milk River and the Missouri.

Meleagris gallopavo.
WILD TURKEY.

Examples of this species are seen occasionally in Southern Ontario. Mr. SAUNDERS reports a pair nesting in Middlesex County, in 1878, and Dr. GARNIER took one in Kent County, in 1884.

Columba fasciata.
BAND-TAILED PIGEON.

A common summer resident of British Columbia.

Ectopistes migratorius.
PASSENGER PIGEON.

The "Wild Pigeon," once so very abundant in all the Eastern Provinces, is now very rare east of Manitoba. It is abundant on the Plains, and rare in British Columbia. Prof. MACOUN met with it in the Saskatchewan Valley; Dr. BELL found it at York Factory, and DREXLER took one example at Moose Factory.

Zenaidura macroura.
MOURNING DOVE.

This species occurs sparingly from the Atlantic to the Pacific, being most frequently met with in Southern Ontario, and in the Red River Valley. COMEAU has taken three specimens at Point des Monts, P. Q.

Columbigallina passerina.
GROUND DOVE.

THOMPSON reports that HIND claims to have handled a specimen that was taken near Winnipeg.

Pseudogryphus californianus.
CALIFORNIA VULTURE.

This species has been taken at the mouth of the Fraser River, B. C.

Cathartes aura.
TURKEY VULTURE.

This species is abundant on the Plains, and fairly common in the southern portions of British Columbia. It occurs regularly at St. Clair Flats, but east of that point is only accidental. A few specimens have been taken at Grand Manan, and Mr. PHILIP COX lately reported the occurrence of two at the mouth of the Miramichi River, in the Gulf of St. Lawrence.

Catharista atrata.
BLACK VULTURE.

A few examples have been taken at Grand Manan.

Elanoides forficatus.
SWALLOW-TAILED KITE.

One example of this species has been seen at Ottawa (*O. F. N. Club*), and a pair were observed near London (*Saunders*), while several have been noted by Mr. HUNTER in different parts of Manitoba (*Thompson*).

Circus hudsonius.
MARSH HAWK.

A common summer resident throughout the Dominion, north to the Fur Countries.

Accipiter velox.
SHARP-SHINNED HAWK.

A common summer resident throughout the southern portions of the Dominion.

Accipiter cooperi.
COOPER'S HAWK.

A summer resident throughout the Dominion to the Fur Countries, but nowhere common.

Accipiter atricapillus.
AMERICAN GOSHAWK.

Occurs between Western Manitoba and the Atlantic, and north to the Fur Countries, though rare in the Huron-Superior region. TURNER considers it as sedentary in the Ungava district; THOMPSON thinks it leaves Manitoba during the winter months; it occurs about the Great Lakes only in winter, and is found in New Brunswick during the entire year.

Accipiter atricapillus striatulus.
WESTERN GOSHAWK.

Occurs in British Columbia, north to Alaska.

Buteo borealis.
RED-TAILED HAWK.

Occurs from the Atlantic to the Great Plains as a summer resident. A few winter in Southern Ontario. On the sea-board it has not been taken north of latitude 49°; but Dr. BELL reports it at Fort Churchill, on Hudson's Bay.

Buteo borealis calurus.
WESTERN RED-TAIL.

Occurs in British Columbia, from the Rocky Mountains to the Pacific.

Buteo lineatus.
RED-SHOULDERED HAWK.

A rather common summer resident of the Eastern Provinces, probably more abundant in Ontario than elsewhere. Dr. BELL reports its occurrence at York Factory, on Hudson's Bay; and Mr. THOMPSON, on the authority of Mr. HUNTER, gives it as rather common in Eastern Manitoba.

Buteo swainsoni.
SWAINSON'S HAWK.

This is an abundant summer resident in Manitoba, and occurs westward to the Cascade Mountains; Mr. McILWRAITH considers it a regular, though rare, visitor to Ontairo; Dr. HALL considered it rare near Montreal in his day—some fifty years ago, and in the DUNLAP-WINTLE list but two specimens are recorded.

Buteo latissimus.
BROAD-WINGED HAWK.

A common summer resident from the Atlantic to Manitoba, ranging west into Assiniboia and north to the Saskatchewan, though on the coast it is not found north of New Brunswick. MCILWRAITH gives it as only a casual visitor to Ontario, but THOMPSON reports it as very abundant in the Muskoka district, and SCOTT considers it one of the commonest Hawks in the Ottawa Valley.

Archibuteo lagopus sancti-johannis.
AMERICAN ROUGH-LEGGED HAWK.

This sub-species occurs throughout Canada, but is of rather irregular distribution. It is fairly common in British Columbia, and is a common autumn migrant through Manitoba and Ontario and part of Quebec. TURNER reports it as more abundant on the eastern and northern shores of Labrador than on the southern, and as breeding at Fort Chimo. COMEAU reports it as rather common, and breeding, at Point des Monts. It occurs in the Maritime Provinces in winter only.

Archibuteo ferrugineus.
FERRUGINOUS ROUGH-LEG.

Dr. COUES found this species breeding among the Pembina hills; Professor MACOUN met with it at Strong Current Creek, and Captain BLAKISTON, along the Saskatchewan Valley.

Aquila chrysaëtos.
GOLDEN EAGLE.

This species occurs throughout Canada, but is rare except in a few localities. Mr. FANNIN reports it as abundant in the Simil-

H

kameen district, though rare along the Pacific coast. Dr. MERRIAM reports that COMEAU has found it breeding at Point des Monts, and "not particularly rare." TURNER found it breeding in the Ungava district, and COUES discovered one pair on the Sweet Grass Hills. It ranges from the far north to Mexico, and is probably sedentary throughout.

Haliæetus leucocephalus.
BALD EAGLE.

A common bird throughout the Dominion, being, probably, most abundant in British Columbia, and rarest along the Great Lakes. Breeds from our southern border to the Arctics.

Falco islandus.
WHITE GYRFALCON.

This is an Arctic bird, but in winter it occasionally wanders as far south as the Bay of Fundy.

Falco rusticolus.
GRAY GYRFALCON.

TURNER reports that this species appeared at Fort Chimo in winter only; he could not learn of any breeding in the Ungava district.

Mr. BOARDMAN has two specimens that were taken in the Bay of Fundy.

Falco rusticolus gyrfalco.
GYRFALCON.

This species occurs from the Hudson's Bay district westward, and north to the Arctic Ocean.

Falco rusticolus obsoletus.
BLACK GYRFALCON.

TURNER reports this species as breeding abundantly at Fort Chimo, though rarely seen there in the winter.

Falco peregrinus anatum.
DUCK HAWK.

This species occurs throughout the entire Dominion, breeding usually in the northern portions, though Dr. COUES found a pair nesting near Milk River, and it has been observed in the breeding season in New Brunswick. It is more common along the ocean borders than in the interior.

Falco peregrinus pealei.
PEALE'S FALCON.

This is reported as a rare bird on Vancouver Island and northward.

Falco columbarius.
PIGEON HAWK.

A common summer resident of Canada at large, breeding usually from about the 45th parallel northward, though THOMPSON reports that in Western Manitoba it occurs as a fall migrant only. It has been taken in New Brunswick in winter, but usually winters in the south.

Falco columbarius suckleyi.
BLACK MERLIN.

A summer resident of British Columbia, though found only between the Cascade Mountains and the Rockies.

Falco richardsonii.
RICHARDSON'S MERLIN.

A rare bird along the Pacific coast and on the Prairies. It has been taken as far east as Sault St. Marie.

Falco sparverius.
AMERICAN SPARROW HAWK.

A common summer resident throughout Canada, from the southern boundary to the Hudson's Bay district.

Pandion haliaëtus carolinensis.
AMERICAN OSPREY.

This is better known to Canadians as the "Fish Hawk." It is a common summer resident of the sea-coasts and the larger lakes north to the Hudson's Bay district and Alaska, and occurs with more or less regularity everywhere but on the Plains. It winters in northern South America.

Strix pratincola.
AMERICAN BARN OWL.

An accidental straggler from the south. Mr. McILWRAITH mentions the occurrence of four examples in Southern Ontario.

Asio wilsonianus.
AMERICAN LONG-EARED OWL.

Occurs in more or less abundance from the Atlantic coast to the Cascade Mountains, and north to the Saskatchewan and the Hudson's Bay district. It is met with in summer, only, in British Columbia and on the Plains, but Mr. NEILSON reports finding it occasionally, in winter, near Quebec, and Mr. DUNLOP has had a similar experience in the vicinity of Montreal.

Asio accipitrinus.
SHORT-EARED OWL.
Occurs throughout the entire Dominion.

Syrnium nebulosum.
BARRED OWL.
Occurs from the southern boundary to about the 50th parallel. Is abundant in the Maritime Provinces, common thence to Manitoba, and rare on the Pacific slope.

Ulula cinerea.
GREAT GRAY OWL.
Occurs in the Arctic regions, and occasionally in winter is found along our southern border, from the Atlantic coast to the Cascades.

Nyctala tengmalmi richardsoni.
RICHARDSON'S OWL.
A northern species, occurring usually in the Maritime Provinces in winter only, though Mr. CORY reports it breeding on the Magdalene Islands. Mr. THOMPSON gives it as a resident in Manitoba.

Nyctala acadica.
SAW-WHET OWL.
A common species throughout Canada to about the 50th parallel.

Megascops asio.
SCREECH OWL.
Occurs from Lake Huron to the Atlantic, though rare in the Maritime Provinces. Prof. MACOUN reports taking one example at Birtle.

Megascops asio kennicottii.
KENNICOTT'S SCREECH OWL.
Very abundant along the coast in British Columbia.

Bubo virginianus.
GREAT HORNED OWL.
Occurs along the Atlantic to Labrador, and westward to the Hudson's Bay district and the Great Plains, though most abundant near the sea-board.

Bubo virginianus subarcticus.
WESTERN HORNED OWL.
Mr. THOMPSON reports that this variety is a common resident of Manitoba.

Bubo virginianus arcticus.
ARCTIC HORNED OWL.
Occurs in the interior from the southern boundary (in winter), to the Fur Countries.

Bubo virginianus saturatus.
DUSKY HORNED OWL.
An abundant resident of British Columbia. TURNER reports it as not rare at Fort Chimo, Labrador.

Nyctea nyctea.
SNOWY OWL.
Occurs throughout the Dominion, though seen in winter only along our southern border. TURNER reports it breeding in Northern Labrador; and Dr. BREWER states, on the authority of Mr. DOWNS, that it has been found breeding in Newfoundland.

Surnia ulula caparoch.
AMERICAN HAWK OWL.

Occurs throughout the Dominion, breeding from the Fur Countries northward, and migrating to our southern border in winter. It is, usually, a rather rare bird everywhere, though common during some seasons.

Speotyto cunicularia hypogæa.
BURROWING OWL.

Mr. FANNIN has found this species only between the Cascade Mountains and the Rockies; Dr. COUES reports it occurring between the Rocky Mountains and Milk River.

Glaucidium gnoma.
PYGMY OWL.

A resident of British Columbia, especially abundant along the coast.

Coccyzus americanus.
YELLOW-BILLED CUCKOO.

Occurs from New Brunswick to British Columbia, but is rare everywhere except in Southern Ontario, and is not mentioned in Mr. THOMPSON's Manitoba list.

Coccyzus erythrophthalmus.
BLACK-BILLED CUCKOO.

A common summer resident from the Maritime Provinces to the Plains; migrating, in winter, south to the West Indies.

Ceryle Alcyon.
BELTED KINGFISHER.

An abundant summer resident throughout Canada to the Fur Countries; migrating, in winter, to the West Indies.

Dryobates villosus leucomelas.
NORTHERN HAIRY WOODPECKER.

An abundant resident from the southern border to the Fur Countries, and west to the Plains (possibly to the Rockies).

Dryobates villosus harrisii.
HARRIS'S WOODPECKER.

An abundant resident of British Columbia; occurs, also, on the eastern slopes of the Rocky Mountains.

Dryobates pubescens.
DOWNY WOODPECKER.

An abundant resident throughout the Dominion (excepting on the Prairies) northward to the lower Fur Countries.

Dryobates pubescens gairdnerii.
GAIRDNER'S WOODPECKER.

Abundant in British Columbia.

Xenopicus albolarvatus.
WHITE-HEADED WOODPECKER.

Occurs in British Columbia, between the Cascade Mountains and the Rockies.

Picoides arcticus.
ARCTIC THREE-TOED WOODPECKER.

This was formerly called the "Black-backed." It is a Northern species, occurring sparingly, in winter, along the southern borders, from the Maritime Provinces westward. THOMPSON gives it in his Manitoban list, but FANNIN does not mention it, though it has been taken in the Cascade Mountains. A few have been taken in New Brunswick in summer.

Picoides americanus.
AMERICAN THREE-TOED WOODPECKER.

This was formerly called the "Banded-backed." It has a more northerly range than *arcticus*, and occurs throughout the Fur Countries to the Arctics, but is seldom seen to the southward, and then only in winter, excepting in the Adirondack region, where Dr. MERRIAM has proved it to be sedentary; and in New Brunswick, where BOARDMAN has taken a few specimens in summer. Mr. BREWSTER reports finding a female with young brood on Anticosti, also.

MCILWRAITH states, on the authority of Mr. TISDALE, that it does not occur in the Muskoka district, but Mr. SCRIVEN reports finding it there, near Gravenhurst. Mr. WHITE, of Ottawa, has a specimen taken in that vicinity on November 5, 1883. Several have been taken in Manitoba.

Sphyrapicus varius.
YELLOW-BELLIED SAPSUCKER.

A common summer resident from the Maritime Provinces to the Prairies, and north to the Saskatchewan. Is rare near Quebec, and COMEAU has taken but one specimen at Point des Monts. Dr. COUES found it plentiful at Pembina, and traced it west to the Souris River. It winters in the far south.

Sphyrapicus varius nuchalis.
RED-NAPED SAPSUCKER.
Occurs in British Columbia, between the Cascade Mountains and the Rockies.

Sphyrapicus ruber.
RED-BREASTED SAPSUCKER.
An abundant resident of British Columbia.

Ceophlœus pileatus.
PILEATED WOODPECKER.
A resident throughout the Dominion in heavily wooded districts, though uncommon or rare east of British Columbia, where it is abundant.

Melanerpes erythrocephalus.
RED-HEADED WOODPECKER.
This species is rare or accidental in the Maritime Provinces and in the vicinity of Quebec, though common near Montreal, throughout Ontario, and in all suitable localities westward to the Rockies. A few have been observed in Ontario during mild winters (*Saunders*).

Melanerpes torquatus.
LEWIS'S WOODPECKER.
Mr. FANNIN reports that this species has been found only between the Cascade Mountains and the Rockies, in British Columbia. Dr. COUES took one specimen on the eastern foothills of the Rockies, near latitude 49°.

Melanerpes carolinus.
RED-BELLIED WOODPECKER.

Mr. McIlwraith states that this species is becoming more common in Ontario; Mr. Schoenan reports that it is quite common in Bruce County. It has not been taken in any other part of Canada.

Colaptes auratus.
FLICKER.

This species was formerly called the "Golden-winged woodpecker." It is a summer resident throughout the Dominion, north to the Hudson's Bay district, though less abundant in British Columbia than to the eastward.

Colaptes cafer.
RED-SHAFTED FLICKER.

A resident of British Columbia; very abundant along the coast.

Colaptes cafer saturatior.
NORTHWESTERN FLICKER.

Mr. Fannin reports that he knows of only two specimens having been seen in British Columbia, one at Burrard Inlet, and one on Vancouver Island.

Colaptes chrysoides.
GILDED FLICKER.

A few examples of this southern bird have been taken in British Columbia (*Fannin*).

Antrostomus vociferus.
WHIP-POOR-WILL.

A summer resident from the Maritime Provinces to the Prairies. Dr. BELL reports that he met with no example north of Norway House.

Chordeiles virginianus.
NIGHTHAWK.

An abundant summer resident from the Atlantic coast to Manitoba, and north to the Fur Countries.

Chordeiles virginianus henryi.
WESTERN NIGHTHAWK.

Mr. THOMPSON reports this variety very abundant in Manitoba, and Mr. FANNIN has taken it in British Columbia.

Cypseloides niger.
BLACK SWIFT.

A fairly common bird on Vancouver Island and along the lower Fraser River.

Chætura pelagica.
CHIMNEY SWIFT.

This species is sometimes called the "Chimney Swallow." It is a common summer resident from the Maritime Provinces to the Plains; ranging north to about the 50th parallel near the Atlantic, and farther north in the western portion of its habitat.

Chætura vauxii.
VAUX'S SWIFT.
A rare summer resident of British Columbia.

Trochilus colubris.
RUBY-THROATED HUMMINGBIRD.
Occurs from the Maritime Provinces to the valley of the Red River, breeding in abundance throughout, excepting in Southern Ontario, where it is chiefly a migrant. It has been taken along the Saskatchewan Valley; Mr. THOMPSON reports having seen one specimen at Bracebridge, in the northern part of Ontario; Mr. NEILSON has taken several nests in the Laurentides back of Quebec, several miles from any settlement; and TURNER met with one example at Davis Inlet. It winters south to Cuba.

Trochilus alexandri.
BLACK-CHINNED HUMMINGBIRD.
A fairly common summer resident of British Columbia.

Trochilus rufus.
RUFOUS HUMMINGBIRD.
An abundant summer resident of British Columbia, widely distributed, and occurring north to Sitka.

Trochilus alleni.
ALLEN'S HUMMINGBIRD.
A summer resident of British Columbia, between the Cascade Mountains and the sea-coast.

Trochilus calliope.
CALLIOPE HUMMINGBIRD.

A fairly common summer resident of British Columbia between the Cascade Mountains and the sea-coast.

Milvulus forficatus.
SCISSOR-TAILED FLYCATCHER.

The general habitat of this species is between Texas and Central America, but a few specimens have wandered into Canada. Dr. BELL reports finding one at York Factory, on Hudson's Bay, in 1880, and of hearing that examples have been seen occasionally at the posts of the Hudson's Bay Company west to the valley of the Mackenzie River; Mr. THOMPSON examined a specimen found by Mr. NASH near Portage la Prairie, in October, 1884; and Dr. GARNIER reports having discovered one in Bruce County, Ontario.

Tyrannus tyrannus.
KINGBIRD.

A common summer resident between the Maritime Provinces and British Columbia (*Fannin*), north to about the 50th parallel. It winters in Central and South America.

Tyrannus verticalis.
ARKANSAS KINGBIRD.

A summer resident from the Plains to the Cascade Mountains.

Myiarchus crinitus.
CRESTED FLYCATCHER.

A common summer resident of Southern Ontario and eastward to Montreal; rare or accidental near Quebec and in New Brunswick, and rare, also, in the Red River Valley.

Sayornis phœbe.
PHŒBE.

A summer resident from the Maritime Provinces to Manitoba, though rare in the latter Province. Winters from the Southern States southward.

Sayornis saya.
SAY'S PHŒBE.

A summer resident of the Prairies, and thence to the Pacific.

Contopus borealis.
OLIVE-SIDED FLYCATCHER.

A summer resident from the Atlantic to the Pacific, and north to about the 50th parallel, though rare in the Great-Lake district. Winters south to Central America.

Contopus virens.
WOOD PEWEE.

A common summer resident from the Maritime Provinces to the eastern edge of the Great Plains.

Contopus richardsonii.
WESTERN WOOD PEWEE.

A common summer resident from Manitoba to the Pacific coast.

Empidonax flaviventris.
YELLOW-BELLIED FLYCATCHER.

A common summer resident of the Maritime Provinces, and rare or casual westward to Manitoba. Mr. COMEAU has taken two specimens at Point des Monts, and Mr. Jos. MACOUN reports finding numbers at Lake Misstassini. KUMLIEN met with one off Cape Farewell, Greenland.

Empidonax difficilis.
Western ~~BAIRD'S~~ FLYCATCHER.

Occurs from the Plains to the Pacific border.

Empidonax acadicus.
ACADIAN FLYCATCHER.

This species is a fairly common summer resident of British Columbia, and Mr. THOMPSON found it breeding at Duck Mountain, Manitoba, in 1884, and not uncommon. Mr. JOHN A. MORDEN reports having taken one example at Hyde Park, Ontario, and Dr. HALL gives it as "scarce" at Montreal.

Empidonax pusillus.
LITTLE FLYCATCHER.

This species is a common summer resident of British Columbia, and Prof. MACOUN found it at the elbow of the South Saskatchewan. It ranges north to the Fur Countries, and south, in winter, to Mexico.

Empidonax pusillus traillii.

TRAILL'S FLYCATCHER.

This species is a fairly common summer resident of the Maritime Provinces, and is usually rare or casual westward to the eastern edge of the Plains, though Dr. Coues reports finding it common near Pembina. Mr. Comeau has taken one specimen at Point des Monts, and in the west it has been taken north to Fort Resolution.

Empidonax minimus.

LEAST FLYCATCHER.

A common summer resident from the Atlantic to the Prairies, though not observed west of Turtle Mountain. On the Atlantic it has not been taken north of Point des Monts, but in the west it ranges farther northward — to Fort Resolution.

Empidonax hammondi.

HAMMOND'S FLYCATCHER.

Dr. Coues took this species on the eastern slope of the Rockies, near the 49th parallel, and Dr. Richardson found it at Lesser Slave Lake.

Empidonax obscurus.

WRIGHT'S FLYCATCHER.

Dr. Coues took this species on the eastern slope of the Rockies, near the 49th parallel.

Otocoris alpestris.
HORNED LARK.

Formerly called the "Shore Lark." It is a common winter visitor to the Maritime Provinces, and occurs westward to Lake Huron. It breeds in the Fur Countries, and northward.

Otocoris alpestris leucolæma.
PALLID HORNED LARK.

Occurs in the western portions of the interior; abundant in Manitoba (*Thompson*).

Otocoris alpestris praticola.
PRAIRIE HORNED LARK.

This form occurs as a summer resident from Montreal to the western edge of the Plains.

Otocoris alpestris arenicola.
DESERT HORNED LARK.

This form occurs as a summer resident of British Columbia, between the Cascade Mountains and the Rockies.

Otocoris alpestris strigata.
STREAKED HORNED LARK.

This form occurs along the coast of British Columbia, and north to Fort Simpson.

Pica pica hudsonica.
AMERICAN MAGPIE.

An abundant resident of British Columbia, east of the Cascades, occurring along the coast in winter only. It occurs on the Plains and in the northern portions of Manitoba, and one example has been taken in Montreal.

Cyanocitta cristata.
BLUE JAY.

A common resident on the Atlantic border to about the 50th parallel, and westward to Manitoba. It has been taken at Moose Factory.

Cyanocitta stelleri.
STELLER'S JAY.

A resident of British Columbia; especially abundant near the coast.

Cyanocitta stelleri macrolopha.
LONG-CRESTED JAY.

A resident of British Columbia; especially abundant near the coast.

Perisoreus canadensis.
CANADA JAY.

This species is known as "Moose Bird" and "Whiskey Jack" in the lumber districts, where it is chiefly found. It is an abundant resident on the Atlantic border north to Labrador, and occurs westward to Manitoba, where it is also abundant. It is absent from Southern Ontario, but Mr. CLEMENTI has taken it near Peterboro', and it is common near Ottawa, and in the Muskoka district.

Perisoreus canadensis capitalis.
ROCKY MOUNTAIN JAY.

Dr. Coues found this form common in the Rockies, near the 49th parallel.

Perisoreus canadensis fumifrons.
ALASKAN JAY.

A resident of British Columbia, east of the Cascade Mountains.

Perisoreus canadensis nigricapillus.
LABRADOR JAY.

Turner found this form in the interior of Labrador, and along the coast.

Perisoreus obscurus.
OREGON JAY.

Occurs on the islands and mainland along the Pacific coast of British Columbia.

Corvus corax sinuatus.
AMERICAN RAVEN.

Occurs throughout the entire Dominion, though of rather local distribution.

Corvus americanus.
AMERICAN CROW.

Occurs throughout the Dominion, north to the Fur Countries.

Corvus caurinus.
NORTHWEST CROW.
A very abundant resident along the coast of British Columbia.

Picicorvus columbianus.
CLARKE'S NUTCRACKER.
This species is abundant in the Similkameen Valley, B. C., and occurs, occasionally, westward to Vancouver Island.

Cyanocephalus cyanocephalus.
PIÑON JAY.
This species occurs in British Columbia.

Dolichonyx oryzivorus.
BOBOLINK.
A common summer resident from the Maritime Provinces to Manitoba.

Dolichonyx oryzivorus albinucha.
WESTERN BOBOLINK.
This form occurs on the Western Prairies.

Molothrus ater.
COWBIRD.
A summer resident from the Maritime Provinces to the Cascades, and north to about the 50th parallel; most abundant in Ontario and Manitoba.

Xanthocephalus xanthocephalus.
YELLOW-HEADED BLACKBIRD.

An abundant summer resident of the 'Great Plains. Mr. THOMPSON gives it as a rare straggler to the vicinity of Toronto; and Dr. MERRIAM states that Mr. COMEAU took one example at Point des Monts in September, 1878.

Agelaius phœniceus.
RED-WINGED BLACKBIRD.

An abundant summer resident from the Maritime Provinces to the Great Plains; north to latitude 50° on the Atlantic border, and to latitude 57° in the west. Said to have been taken on the Pacific Slope.

Agelaius gubernator.
BI-COLORED BLACKBIRD.

An abundant summer resident in the southern portion of British Columbia.

Sturnella magna.
MEADOWLARK.

A common summer resident of Southern Ontario, occurring sparingly eastward to the Maritime Provinces, and westward to Eastern Manitoba.

Sturnella magna neglecta.
WESTERN MEADOWLARK.

Occurs from Manitoba to the Pacific, and north to the valley of the Saskatchewan, being most abundant on the Great Plains.

Icterus spurius.
ORCHARD ORIOLE.

Occurs regularly in Southern Ontario, and Mr. THOMPSON reports that one example was taken at Leslieville in 1879. Dr. COUES secured one at Pembina, and several have been taken at St. Stephen, N. B.

Icterus galbula.
BALTIMORE ORIOLE.

This species is an abundant summer resident of the Great Plains; is common in Southern Ontario, and rare or casual eastward to the Maritime Provinces. Capt. BLAKISTON found it in the valley of the Saskatchewan, about latitude 55°, and Dr. BELL has the skin of a specimen taken by Dr. MATTHEW at York Factory.

Icterus bullocki.
BULLOCK'S ORIOLE.

Mr. FANNIN has found this species between the Cascade Mountains and the Rockies, in British Columbia.

Scolecophagus carolinus.
RUSTY BLACKBIRD.

Occurs throughout Canada, breeding from about the 45th parallel to the Fur countries; enormously abundant in Manitoba (*Thompson*).

Scolecophagus cyanocephalus.
BREWER'S BLACKBIRD.

An abundant summer resident from Manitoba to the Cascades.

Quiscalus quiscalus æneus.
BRONZED GRACKLE.

An abundant summer resident from the Maritime Provinces to the Great Plains. Dr. BELL reports it at York Factory, H. B. T.

Coccothraustes vespertina.
EVENING GROSBEAK.

An abundant resident of British Columbia, east of the Cascades, and occasionally found on the western slope and in Vancouver Island. It is a common winter visitor to Manitoba, and a few specimens have been taken in Ontario.

Pinicola enucleator.
PINE GROSBEAK.

Occurs throughout Canada, breeding from the Fur countries northward; in winter it is found from about the 50th parallel southward. Mr. COMEAU finds it throughout the year at Point des Monts, and Mr. SCRIVEN has found a few in summer in the Muskoka district; it is casual in New Brunswick, also, in summer. In British Columbia it is more common on the eastern side of the Cascades than on the Pacific Slope, and is occasional, only, on Vancouver Island.

Carpodacus purpureus.
PURPLE FINCH.

An abundant bird from the Atlantic border to the Prairies, and north to about the 50th parallel. It winters sparingly in Southern Ontario, and casually in New Brunswick, and occurs southward to the Southern States.

Carpodacus purpureus californicus.
CALIFORNIA PURPLE FINCH.
An abundant summer resident of British Columbia.

Carpodacus cassini.
CASSIN'S PURPLE FINCH.
Occurs in British Columbia; most common east of the Cascades.

Loxia curvirostra minor.
AMERICAN CROSSBILL.
A common resident throughout Canada, north to about the 50th parallel on the Atlantic border, and ranging farther north in the west. It is a winter visitor, only, in Southern Ontario and Southern Manitoba. Dr. BELL reports taking one specimen in Hudson's Straits, and it has been taken in Sitka.

Loxia leucoptera.
WHITE-WINGED CROSSBILL.
A common species throughout Canada, from the Atlantic to the Cascades, and north to the Arctic regions; but is met with in winter, only, along the southern border, and is somewhat rare in parts of Ontario.

Leucosticte tephrocotis.
GRAY-CROWNED LEUCOSTICTE.
Dr. COUES found this species in the Red River Valley, and it has been taken on the Saskatchewan.

L

Leucosticte tephrocotis littoralis.
HEPBURN'S LEUCOSTICTE.

Mr. FANNIN has found this species in the mountain districts of British Columbia, and occasionally along the coast.

Acanthis hornemannii.
GREENLAND REDPOLL.

Occurs in the Eastern Arctic regions.

Acanthis hornemannii exilipes.
HOARY REDPOLL.

Occurs in the Arctic regions; south, in the fall, to Manitoba (*Thompson*).

Acanthis linaria.
REDPOLL.

Occurs from the Fur Countries to the southern border (in winter), and from the Atlantic to the Plains.

Acanthis linaria holbœllii.
HOLBŒLL'S REDPOLL.

Occurs along the Atlantic border.

Acanthis linaria rostrata.
GREATER REDPOLL.

Occurs from sub-Arctic regions to the southern border, and from the Atlantic westward, probably to Manitoba.

Spinus tristis.
AMERICAN GOLDFINCH.

This species is better known as the "Thistle Finch," and in some localities is called "Wild Canary," a name which is also applied to the Yellow Warbler. It is an abundant summer resident throughout Canada, north to Southern Labrador and the Valley of the Saskatchewan.

Spinus pinus.
PINE SISKIN.

A common summer resident of British Columbia, and a winter visitor of the southern portions of Ontario and Quebec, occurring, also, as a migrant, in Manitoba and the Maritime Provinces. A few summer in New Brunswick.

Passer domesticus.
EUROPEAN HOUSE SPARROW.

An introduced species which has become thoroughly naturalized, and is now found in almost all the towns between Halifax, N. S., and Windsor, Ontario.

Plectrophenax nivalis.
SNOWFLAKE.

This species was formerly called the "Snow Bunting." It occurs abundantly throughout the entire Dominion, breeding in the Arctic regions and wintering along the southern border, and south to Georgia and Kansas.

Calcarius lapponicus.
LAPLAND LONGSPUR.

Occurs from the Atlantic to the Great Plains (probably to the Rockies), and north to sub-Arctic regions; in winter, only, along the southern border, and south to South Carolina and Kansas.

Calcarius pictus.
SMITH'S LONGSPUR.
. Occurs as a migrant on the Prairies, breeding in the far north.

Calcarius ornatus.
CHESTNUT-COLLARED LONGSPUR.
A common summer resident from Manitoba to the foot-hills of the Rockies, and north to the Saskatchewan.

Rhynchophanes mccownii.
McCOUN'S LONGSPUR.
Occurs on the Great Plains, from the Missouri Basin to the Rockies.

Poocætes gramineus.
VESPER SPARROW.
This is the "Grass Finch" of the older authors. It is a common summer resident from the Maritime Provinces to the Great Plains, and occurs north to the mouth of the St. Lawrence, and to the Saskatchewan Valley.

Poocætes gramineus confinis.
WESTERN VESPER SPARROW.
A very abundant summer resident of Manitoba, and occurs thence to the Pacific coast, along which it is rather uncommon.

Ammodramus princeps.
IPSWICH SPARROW.

This species has been taken in New Brunswick and Prince Edward Island during the spring migration, and has been found breeding on Sable Island, off the Atlantic coast of Nova Scotia.

Ammodramus sandwichensis.
SANDWICH SPARROW.

This species is fairly common along the coast of British Columbia.

Ammodramus sandwichensis savanna.
SAVANNA SPARROW.

An abundant summer resident from the Maritime Provinces to the Prairies, and north to the Hudson's Bay region.

Ammodramus sandwichensis alaudinus.
WESTERN SAVANNA SPARROW.

A very abundant summer resident of Manitoba, and thence to the Pacific coast.

Ammodramus bairdii.
BAIRD'S SPARROW.

Occurs on the Great Plains.

Ammodramus savannarum passerinus.
GRASSHOPPER SPARROW.

Mr. SAUNDERS considers this a fairly common resident in the southwestern Peninsula of Ontario. Mr. BOARDMAN has taken one example near St. Stephen, N. B.

Ammodramus henslowii.
HENSLOW'S SPARROW.
This species is said to occur in Ontario (*A. O. U. Check-List*).

Ammodramus leconteii.
LECONTE'S SPARROW.
A common summer resident in Manitoba and Assiniboia.

Ammodramus caudacutus.
SHARP-TAILED SPARROW.
With the exception of one example, reported by the OTTAWA FIELD NATURALISTS' CLUB as having been taken in their vicinity, there is no record of this bird's occurrence outside of New Brunswick and Prince Edward Island. It has been found more abundantly on the Marshes at the head of the Bay of Fundy, though it is common near St. Andrews, and has been taken above St. Stephen and on the marshes near Hampton. Messrs. STONE and BREWSTER report its occurrence on Prince Edward Island.

Chondestes grammacus strigatus.
WESTERN LARK SPARROW.
A common summer resident in British Columbia and on the Great Plains, and casual or rare eastward to Toronto.

Zonotrichia querula.
HARRIS'S SPARROW.
An abundant migrant on the Great Plains.

Zonotrichia leucophrys.
WHITE-CROWNED SPARROW.

A summer resident of the Atlantic border, north to Labrador, though of local distribution. It is reported as a rare migrant near Montreal, in Manitoba, and on the Prairies, but is fairly common in portions of Ontario. Mr. FANNIN has not mentioned it in his list of birds of British Columbia, though it is said to occur in the Rocky Mountain region in abundance.

Zonotrichia intermedia.
INTERMEDIATE SPARROW.

A summer resident of British Columbia, and found by Dr. COUES migrating along the Souris River.

Zonotrichia gambeli.
GAMBEL'S SPARROW.

A summer resident along the Pacific coast; very abundant near Victory.

Zonotrichia coronata.
GOLDEN-CROWNED SPARROW.

A summer resident of British Columbia; abundant on Vancouver Island.

Zonotrichia albicollis.
WHITE-THROATED SPARROW.

This species is better known to Canadians as the "Kennedy Bird," or "Peabody Bird," or "Old Tom Peabody." It is an abundant summer resident from the Atlantic to the Prairies, and north to the Fur Countries.

Spizella monticola.
TREE SPARROW.
A common bird from the Atlantic to the Rockies; breeding in the Fur Countries, and wintering along the southern border, and southward to the Southern States.

Spizella monticola ochracea.
WESTERN TREE SPARROW.
A fairly common summer resident of British Columbia.

Spizella socialis.
CHIPPING SPARROW.
Occurs from the Atlantic border to the Rockies, though rare on the Prairies. It occurs north to Gaspé (*Brewster*) in the east, and to Great Slave Lake in the west — breeding from our Southern border northward. South, in winter, to Eastern Mexico.

Spizella socialis arizonæ.
WESTERN CHIPPING SPARROW.
A common summer resident of British Columbia.

Spizella pallida.
CLAY-COLORED SPARROW.
A very abundant resident of Manitoba, and occurs across the Plains to the Rockies.

Spizella breweri.
BREWER'S SPARROW.
A fairly common summer resident of British Columbia.

Spizella pusilla.
FIELD SPARROW.

A fairly common summer resident in the vicinity of Quebec and Montreal, and occurs thence westward to Manitoba.

Junco hyemalis.
SLATE-COLORED JUNCO.

This species is the "Snowbird" of the older authors. It occurs in abundance from the Atlantic coast to Lake Superior, and north to the Fur Countries; breeding throughout, though sparingly in the more southern portions of Ontario and Quebec, where, also, a few winter. It has also been observed in winter in New Brunswick. It is said to occur on the Pacific, but is not mentioned by Mr. THOMPSON in his Manitoba list.

Junco hyemalis oregonus.
OREGON JUNCO.

This form is a very abundant resident of British Columbia.

Peucæa ruficeps.
RUFOUS-CROWNED SPARROW.

A summer resident along the Pacific coast; very abundant on Vancouver Island.

Melospiza fasciata.
SONG SPARROW.

An abundant summer resident from the Atlantic coast to the wooded districts of the Plains, and northward to the lower Fur Countries. It winters sparingly along the southern border.

M

Melospiza fasciata guttata.
RUSTY SONG SPARROW.
A fairly abundant resident near the Pacific coast (*Fannin*).

Melospiza fasciata rufina.
SOOTY SONG SPARROW.
An abundant resident along the sea-board of British Columbia.

Melospiza lincolni.
LINCOLN'S SPARROW.
Occurs throughout Canada, north to the lower Fur Countries, but is most numerous in the mountain regions of British Columbia. Very few examples have been observed in Ontario, though Dr. WHEATON mentions that it is a "not uncommon" migrant near Sandusky and Cleveland. It has been taken in a few localities in the Maritime Provinces.

Melospiza georgiana.
SWAMP SPARROW.
A common summer resident from the Atlantic to the Plains; breeding from the southern border northward. AUDUBON found it in Labrador, and RICHARDSON mentions its occurrence at Fort Simpson.

Passerella iliaca.
FOX SPARROW.
Occurs throughout Canada, though of rather local distribution. It is abundant in portions of the Maritime Provinces, and near Quebec as a spring migrant, though rarely met with in the autumn. Mr. MCILWRAITH mentions that it is seldom seen near Hamilton, and it is generally considered. rare through-

out Ontario, as well as in the vicinity of Ottawa and Montreal, though Mr. CLEMENTI thinks it fairly common near Peterboro, and Dr. WHEATON reports that it is common in Ohio. It is common in Manitoba, and uncommon in British Columbia; and, while it has not been reported from the Hudson's Bay region, is said to extend its range to the Arctic coast. It has been found breeding in Labrador, and on the islands in the Gulf of St. Lawrence, and on Duck Mountain, Manitoba (*Thompson*), and is a summer resident near the sea-coast in British Columbia (*Fannin*).

Passerella iliaca unalaschcensis.
TOWNSEND'S SPARROW.

A fairly common summer resident along the coast of British Columbia.

Pipilo erythrophthalmus.
TOWHEE.

A common summer resident in Manitoba, the Great Plains, and in Southern Ontario, east to Toronto. Mr. NEILSON reports that a pair were taken near Quebec in 1879, and one example has been taken in New Brunswick, near St. John.

Pipilo maculatus arcticus.
ARCTIC TOWHEE.

Occurs on the Saskatchewan Plains, and west to the eastern slopes of the Rockies.

Pipilo maculatus oregonus.
OREGON TOWHEE.

An abundant resident of British Columbia, from the Cascades to the Pacific.

Cardinalis cardinalis.
CARDINAL.
This species occurs occasionally in Southern Ontario, and two examples were observed near Halifax, N. S., in 1871 (*Jones*).

Habia ludoviciana.
ROSE-BREASTED GROSBEAK.
Occurs sparingly in the Maritime Provinces, but is more common in the vicinity of Quebec, and is fairly common in Ontario (north to Gravenhurst) and throughout Manitoba. Dr. COUES found it in abundance at Pembina, and it has been taken on the banks of the Saskatchewan.

Habia melanocephala.
BLACK-HEADED GROSBEAK.
A summer resident of British Columbia.

Guiraca cærulea.
BLUE GROSBEAK.
A straggler from the south that occurs occasionally in British Columbia; also, Mr. COUPER reports taking one specimen near Quebec, and Mr. BOARDMAN reports that a few have been taken on Grand Manan, in the Bay of Fundy.

Passerina cyanea.
INDIGO BUNTING.
A summer resident from the southern portions of the Maritime Provinces to Southern Ontario. Mr. COMEAU reports taking one example at Point des Monts, in 1884.

Passerina amœna.
LAZULI BUNTING.
A rare summer resident of British Columbia.

Spiza americana.
DICKCISSEL.
This is the "Black-throated Bunting" of the older authors. Mr. SAUNDERS reports finding several examples and a nest at Point Pelee, Ontario, in 1884.

Calamospiza melanocorys.
LARK BUNTING.
Prof. MACOUN reports finding this species on the Western Souris Plains and at the Cypress Hills, and Dr. COUES met with it in the Missouri River region, along the 49th parallel.

Piranga ludoviciana.
LOUISIANA TANAGER.
A fairly common summer resident of British Columbia, and occurs, also, on the Prairies.

Piranga erythromelas.
SCARLET TANAGER.
Occurs from the Maritime Provinces to the Great Plains, and north to Lake Winnipeg.

Piranga rubra.
SUMMER TANAGER.
An accidental straggler from the south. Mr. MCILWRAITH has taken one near Hamilton; two were taken near Montreal,

in 1864, by Mr. WILLIAM HUNTER, and examples have been secured in the same vicinity by Mr. JAMES FOLEY and Mr. KEUTZING. One has been sent to Mr. BOARDMAN from Halifax, and two from Grand Manan.

Progne subis.
PURPLE MARTIN.

Occurs from the Maritime Provinces to British Columbia, north to the Saskatchewan. Winters in Mexico.

Petrochelidon lunifrons.
CLIFF SWALLOW.

Occurs from the Maritime Provinces to British Columbia. The most northerly point on the Atlantic border at which it has been observed is Point des Monts, where Mr. COMEAU discovered a small colony in 1882; but in the west it is said to range to the Arctics. It winters as far south as Brazil and Paraguay.

Chelidon erythrogaster.
BARN SWALLOW.

Occurs from the Atlantic border to the Pacific, north to the Fur Countries, and south, in winter, to South America.

Tachycineta bicolor.
TREE SWALLOW.

Formerly called the "White-bellied Swallow." It occurs from the Atlantic border to the Pacific, north to the Fur Countries, and south, in winter, to Central America.

Tachcineta thalassina.
VIOLET-GREEN SWALLOW.

Mr. FANNIN considers this an abundant summer resident of British Columbia. Dr. COUES reports taking one example in the Quaking-Ash River, and Prof. MACOUN secured one on the Wait-a-bit River. It winters south to Guatemala.

Clivicola riparia.
BANK SWALLOW.

Occurs from the Maritime Provinces to the Pacific coast. Professor VERRILL found it common on Anticosti, and Mr. BREWSTER discovered a colony at Gaspé, and another on the Magdalene Islands, where Mr. CORY also found a colony. Mr. COMEAU has seen only one example at Point des Monts. It is said to occur in the Mackenzie River, and to migrate to South America in winter.

Stelgidopteryx serripennis.
ROUGH-WINGED SWALLOW.

This is a common summer resident of British Columbia. Mr. SAUNDERS reports finding it common in the vicinity of London, Ontario, and Rev. Mr. CLEMENTI has found it near Peterboro. It migrates south to Guatemala in winter.

Ampelis garrulus.
BOHEMIAN WAXWING.

This species occurs as a resident in British Columbia, between the Cascade Mountains and the Rockies, and is occasionally seen on Vancouver Island. On the Plains it is abundant, but usually occurs as a winter visitant only, though Prof. MACOUN found a pair near Silver City in summer, and Dr. COUES saw fledglings near Chief Mountain Lake in August. From Manitoba to the Atlantic it occurs irregularly during the winter months.

Ampelis cedrorum.
CEDAR WAXWING.

An abundant summer resident from the Atlantic to the Pacific, north to the Fur Countries; migrating, in winter, to Guatemala.

Lanius borealis.
NORTHERN SHRIKE.

Occurs throughout Canada, breeding far north, and migrating to the southern border in winter.

Lanius ludovicianus excubitorides.
WHITE-RUMPED SHRIKE.

Occurs from the Maritime Provinces (sparingly) to the Rockies, being most abundant on the Prairies.

Vireo olivaceus.
RED-EYED VIREO.

This species is a common summer resident from the Maritime Provinces to southern British Columbia (*Fannin*), and north to Anticosti (*Verrill*), and Fort Simpson (*Richardson*).

Vireo flavoviridis.
YELLOW-GREEN VIREO.

The general habitat of this bird is between Texas and Panama, but one specimen was captured by Mr. COMEAU at Point des Monts, P. Q., on May 13, 1883 (*Merriam*).

Vireo philadelphicus.
PHILADELPHIA VIREO.
In the *A. O. U. Check-List* the habitat of this species is given as "Eastern North America, north to Hudson's Bay"; but, so far, it appears to have escaped detection by the majority of Canadian observers. The records are: "A regular summer resident of Western Manitoba" (*Thompson*); "One specimen taken June 2, 1883, and one September 4, 1884, near Ottawa, by Master TED WHITE" (*Ottawa F. N. Club*); "Several were secured by our party at Edmundston, N. B., in June, 1882" (*Chamberlain*); "Occurs occasionally in the vicinity of St. Stephen, N. B." (*Boardman*). Mr. McILWRAITH also reports that it occurs as a migrant in Ontario, and DREXLER took one at Moose Factory, H. B. T.

Vireo gilvus.
WARBLING VIREO.
A summer resident throughout Canada, to the Fur Countries.

Vireo flavifrons.
YELLOW-THROATED VIREO.
A summer resident of Southern Ontario and Manitoba, and placed by Dr. HALL, and by DUNLOP and WINTLE in their Montreal lists.

Vireo solitarius.
BLUE-HEADED VIREO.
Occurs from the Maritime Provinces to the Great Plains, though generally rather rare, excepting in Manitoba.

Vireo solitarius cassinii.
CASSIN'S VIREO.
A fairly common bird in British Columbia.

N

Vireo noveboracensis.
WHITE-EYED VIREO.

The claim of this species to a place among the birds of Canada is based on the capture of one example near St. John, N. B., in 1877, by Mr. HAROLD GILBERT.

Mniotilta varia.
BLACK AND WHITE WARBLER.

A common summer resident from the Maritime Provinces to the Plains. It has been reported as a migrant, only, at some stations in the more southern district of Ontario and Quebec, but as it breeds in Ohio (*Wheaton*) it is fair to assume that further investigation will reveal its breeding places along our entire southern border. It is an abundant summer resident of Manitoba, and occurs north to Fort Simpson. Winters in the West Indies.

Protonotaria citrea.
PROTHONOTARY WARBLER.

One specimen of this southern Warbler was taken at St. Stephen, N. B., by Mr. GEORGE A. BOARDMAN, on October 30, 1862, and upon that record alone rests the claim of this species to a place among Canadian birds.

Helminthophila chrysoptera.
GOLDEN-WINGED WARBLER.

Mr. SAUNDERS reports that this species is common near London, Ontario.

Helminthophila ruficapilla.
NASHVILLE WARBLER.

A summer resident from the Maritime Provinces to the Great Plains, and north to Labrador (*Nuttall*), and the northwestern Fur Countries (*Richardson*). It is rare in some districts in Southern Ontario. Winters in Mexico.

Helminthophila celata.
ORANGE-CROWNED WARBLER.

An abundant summer resident of the Prairies, ranging north to Great Slave Lake and the Yukon district; casual in Ontario.

Helminthophila celata lutescens.
LEUTESCENT WARBLER.

An abundant summer resident of British Columbia.

Helminthophila peregrina.
TENNESSEE WARBLER.

The Tennessee Warbler, like many other species of our songsters, is too little known at present for a satisfactory account of its distribution to be obtained. It is a common summer resident of New Brunswick, but its name does not appear in Mr. JONES's catalogue of Nova Scotia birds, though Mr. McKINLAY reports it as occurring near Pictou. Mr. COMEAU considers it tolerably common at Point des Monts, Mr. BREWSTER captured one at Anticosti, and Mr. JOSEPH M. MACOUN found it quite common at Lake Misstassini; but it is not in Mr. COUPER's list of Quebec species, and Mr. NEILSON has taken but one example in that vicinity, though Mr. C. E. DIONNE records the occurrence of a few there in the spring of 1879, while Messrs. DUNLOP and WINTLE have found it fairly common near Montreal in the migrations. From Ontario there are but few reports of its occurrence. In the SAUNDERS-MORDEN list it is given as common at times in the migrations; Mr. THOMPSON calls it a rare spring migrant near Toronto, and the OTTAWA CLUB have met with but one example, though Dr. WHEATON mentions it as a rare and irregular spring migrant (in Ohio), but abundant and regular in the fall. Mr. BARNSTON took it on the north shore of Lake Superior. Dr. COUES found it very abundant in migrations at Pembina; and THOMPSON reports it as a rare summer resident of Western Manitoba. DREXLER took examples at Moose Factory and Fort George, on Hudson's Bay, and Mr. Ross met it at Great Slave Lake. It winters in Central America.

Compsothlypis americana.
PARULA WARBLER.

Occurs from the Maritime Provinces to Lake Huron, breeding from about latitude 45° northward. BREWSTER found it breeding on Anticosti. It winters in the West Indies and Central America.

Dendroica tigrina.
CAPE MAY WARBLER.

The only portions of Canada in which the Cape May appears to be at all common are Manitoba and New Brunswick, though it is not an abundant bird anywhere. Mr. THOMPSON reports that it is "plentiful" along the Red River, and Mr. BOARDMAN writes that "though often rare, it is quite common during some seasons," near St. Stephen, on the Maine border, and it has been taken in several other parts of New Brunswick.

The other notes of its occurrence that I have been enabled to gather are but few. Mr. MCILWRAITH has taken but six examples in Ontario; Messrs. SAUNDERS and MORDEN report one taken at Mitchell's Bay; the OTTAWA F. N. CLUB report a pair captured near their headquarters; Messrs. DUNLOP and WINTLE consider it a rare, though regular, migrant in the vicinity of Montreal, while Mr. NEILSON knows of but one specimen that has been observed near Quebec. DREXLER captured several specimens at Moose Factory, in Hudson's Bay.

It breeds on the hills in Jamaica, and winters in Central America.

Mr. JAMES W. BANKS, of St. John, N. B., has a nest and eggs of this species taken by him near that city, and they remain unique — no other nest and eggs, taken in the United States or Canada, are to be found in any collection.

Dendroica æstiva.
YELLOW WARBLER.

This species, known formerly as the "Summer Yellowbird" and "Summer Warbler," and called by the people of the Maritime Provinces the "Wild Canary," is an abundant summer resident from the Atlantic to the Pacific, and occurs north to Fort George, on Hudson's Bay, and to Fort Simpson, in the Great Slave Lake district. It winters in Central America.

Dendroica cærulescens.
BLACK-THROATED BLUE WARBLER.

The *A. O. U. Check-List* gives the habitat of this species as "Eastern North America to the Plains," but it has not been observed in Canada west of Lake Huron. It occurs in the Maritime Provinces as a rare summer resident, but is more common in Quebec and Ontario, where it has been observed only during the migrations, though it probably breeds in portions of these latter Provinces. It winters in the West Indies.

Dendroica coronata.
MYRTLE WARBLER.

The Myrtle Warbler, heretofore called the "Yellow Rump," occurs throughout Canada. It breeds abundantly in the Maritime

Provinces, and is one of the few of this family that Mr. CORY found breeding in abundance on the Magdalene Islands. Mr. COMEAU has found it rather common at Point des Monts, and Mr. STEARNS records it as common in the interior of Labrador. Several specimens were taken at Moose Factory by DREXLER, and KUMLIEN secured a single example in Godhaven Harbour, Greenland.

To the westward of the Atlantic Provinces it is generally reported as a migrant, though it is a summer resident of British Columbia.

In winter it occurs from the Middle States to Central America.

Dendroica auduboni.
AUDUBON'S WARBLER.

An abundant summer resident of British Columbia, and also abundant on the eastern slope of the Rockies.

Dendroica maculosa.
MAGNOLIA WARBLER.

This species, formerly called the " Black-and-Yellow Warbler," occurs in British Columbia as a rare summer resident, and in Manitoba as a rare migrant; and is, also, a migrant only, though more common, through the settled portions of Ontario, and parts of Quebec; but north and east of these latter districts it is found during the summer, and in the Maritime Provinces it breeds abundantly. It is the commonest Warbler on the north and south shores of the mouth of the St. Lawrence, and on Anticosti, and Mr. JOSEPH MACOUN found it quite common at Lake Misstassini. It has been taken in Labrador and in the Slave Lake district.

Winters in the West Indies and Central America.

Dendroica cærulea.
CERULEAN WARBLER.

Mr. McIlwraith considers this species a regular summer resident in Southern Ontario, though of local distribution. Mr. Couper reports that in the spring of 1866 it was very common near Quebec.

Dendroica pensylvanica.
CHESTNUT-SIDED WARBLER.

A common summer resident from the Maritime Provinces to the Great Plains. South in winter to Central America.

Dendroica castanea.
BAY-BREASTED WARBLER.

Occurs from the Maritime Provinces to the Prairies, being most abundant in the west. Has been taken in Labrador, and on the west coast of Hudson's Bay. Winters in Central America.

Dendroica striata.
BLACK-POLL WARBLER.

Occurs throughout Canada to the Arctics. It is abundant in Labrador and Manitoba, rare in British Columbia and Ontario, and common elsewhere. It breeds from latitude 45° (sparingly) northward. Winters in northern South America.

Dendroica blackburniæ.
BLACKBURNIAN WARBLER.

Occurs from the Atlantic border to the Plains, and north to Labrador; breeding from about the 45th parallel.

Dendroica nigrescens.
BLACK-THROATED GRAY WARBLER.

A fairly common summer resident of British Columbia. Winters in Mexico.

Dendroica virens.
BLACK-THROATED GREEN WARBLER.

Occurs from the Atlantic border to Lake Huron, and north to Point des Monts. It is said to range to the Great Plains, but has not been observed in Manitoba.

Winters south to Cuba and Panama.

Dendroica townsendi.
TOWNSEND'S WARBLER.

A common summer resident of British Columbia. Winters in Guatemala.

Dendroica occidentalis.
HERMIT WARBLER.

A rare summer resident of Vancouver Island. Winters in Guatemala.

Dendroica vigorsii.
PINE WARBLER.

This species occurs from the Atlantic border to the Plains. It is fairly common in Ontario, but rare or casual elsewhere,— the only record for Manitoba being Prof. Macoun's report of finding it in Duck Mountain. Winters in the Southern States.

Dendroica palmarum.
PALM WARBLER.

This species is an abundant migrant in Manitoba, and has been seen as far north as Great Slave Lake. It winters in the Gulf States.

Dendroica palmarum hypochrysea.
YELLOW PALM WARBLER.

This is the eastern form, and is familiarly known by its old name of "Yellow Redpoll." It is an abundant summer resident of New Brunswick, breeding in all suitable localities; but it is not so plentiful in Nova Scotia, and Mr. BAIN thinks it rare in Prince Edward Island. It occurs as a rare or occasional migrant, only, in Quebec and Ontario. It has been taken at Moose Factory and other points about Hudson's Bay, and Mr. COMEAU secured one specimen at Point des Monts, in May, 1885.

It winters in South Atlantic and Gulf States.

Seiurus aurocapilus.
OVEN-BIRD.

A common summer resident from the Maritime Provinces to the Plains, and rare in British Columbia (*Fannin*). Occurs north to Labrador (*Stearns*), the Hudson's Bay region, and Alaska. Winters south to Central America.

Seiurus noveboracensis.
WATER THRUSH.

This Warbler occurs from the Atlantic border to Manitoba, and north to the Fur Countries. It is fairly common in the Maritime Provinces, though somewhat rare near Point des Monts and Quebec; but Mr. STEARNS reports it as not uncommon in the interior of Labrador, and Mr. JOSEPH MACOUN found it common at Lake Misstassini.' Dr. BELL observed it in the Straits of Belle Isle, TURNER took several specimens at Davis Inlet, and one has been taken in Greenland. In Dr. HALL's list of Montreal birds, as well as in that of Messrs. DUNLOP and WINTLE, it is given as rare, and a similar report is made for Ottawa by the FIELD NATURALISTS' CLUB. Mr. SAUNDERS considers it not at all common in Southern Ontario, and Mr. SCHOENAN reports it as rare in Bruce County; but Mr. McILWRAITH, in his *Birds of Ontario*, gives it as quite abundant. It winters south to northern South America.

Seiurus noveboracensis notabilis.
GRINNELL'S WATER THRUSH.

This western sub-species of the Water Thrush is a common summer resident of the Prairies, and is also found in California. Winters south to northern South America.

Seiurus motacilla.
LOUISIANA WATER THRUSH.

This form occurs sparingly along the southern border of Ontario.

Geothlypis formosa.
KENTUCKY WARBLER.

As this bird seldom comes so far north as southern New England, its appearance in Canada must be considered as merely accidental. The only known instance of its occurrence here is that of a pair seen several times by Mr. JOHN NEILSON, near the City of Quebec, during the first half of July, 1879.

Geothlypis agilis.
CONNECTICUT WARBLER.

Very little is yet known of the distribution of this species. In the *A. O. U. Check-List* the habitat is given as "eastern North America, breeding north of the United States." In *New England Bird Life* it is stated that this bird is a migrant through Connecticut and Massachusetts, where it is rare in the spring, though more abundant in the autumn; but it has no New England record beyond these States. So far it has not been observed in any of the Maritime Provinces, nor in Quebec; and it was unknown in Ontario until Mr. W. E. SAUNDERS took one specimen near London, in September, 1883, and since then he has found them fairly common in the same vicinity and at Point Pelee for a few days in June. Its regular occurrence in Ontario was to have been expected, as Dr. WHEATON had reported it as a regular migrant through Northern Ohio. Mr. THOMPSON reports it as a fairly common summer resident in Western Manitoba.

Geothlypis philadelphia.
MOURNING WARBLER.

Dr. COUES found the Mourning Warbler "breeding abundantly" at Pembina, and Mr. THOMPSON reports it as common in Manitoba, and in some localities abundant, but east of that Province it is generally reported as of rare or uncommon occurrence. It is rather rare, usually, in New Brunswick and Nova Scotia, though Mr. FRANCIS BAIN reports it as common on Prince Edward Island. It winters in Central America.

Geothlypis macgillivrayi.
MACGILLIVRAY'S WARBLER.

A fairly common summer resident of British Columbia.

Geothlypis trichas.
MARYLAND YELLOW-THROAT.

An abundant summer resident from the Atlantic border to Lake Huron, though of somewhat local distribution in Ontario. TURNER reports it common in the southern portions of Labrador. Winters south to the Gulf States and West Indies.

Geothlypis trichas occidentalis.
WESTERN YELLOW-THROAT.

An abundant summer resident in Manitoba and west to the Rockies; fairly common in British Columbia. Winters south to Central America.

Icteria virens.
YELLOW-BREASTED CHAT.

A few examples of this species have been observed in Ontario, but its occurrence there is merely accidental. Dr. COUES reports finding it in abundance in the Missiouri Valley region, near the 49th parallel.

Sylvania mitrata.
HOODED WARBLER.

This is a bird of the Carolinian Fauna, but has gained a right to be mentioned here by occasionally straggling into Southern Ontario.

It winters in the West Indies, and in Central America.

Sylvania pusilla.
WILSON'S WARBLER.

"Wilson's Black Cap," also known as the "Black-capped Yellow Warbler," is reported as rare in Ontario, but east and west of that Province it is fairly common, though it has been found breeding only in New Brunswick and Nova Scotia, and, occasionally, in the Fur Countries. It winters in Central America.

Sylvania pusilla pileolata.
PILEOLATED WARBLER.
An abundant summer resident of British Columbia.

Sylvania canadensis.
CANADIAN WARBLER.
The "Canadian Flycatcher," or "Flycatching Warbler," as this species has been called, is a more or less common species from the Atlantic border to the Great Plains, ranging as far north as Lake Winnipeg and Point des Monts. It is a migrant in the southern portions of Manitoba and Ontairo, but elsewhere is a summer resident.

Winters south to northern South America.

Setophaga ruticilla.
AMERICAN REDSTART.
This Warbler occurs from the Atlantic border to the Pacific, and north to Labrador and Fort Simpson, breeding throughout its range. It is very abundant in many localities in the Maritime Provinces and in Manitoba, though rare in British Columbia. Winters from Southern Mexico to northern South America.

Anthus pensylvanicus.
AMERICAN PIPIT.
This species, formerly called the "Titlark," occurs in more or less abundance throughout the Dominion, but is only met with

along the southern portions while migrating between its winter home in the south and its breeding grounds in sub-arctic regions. KUMLIEN found it breeding as far north as latitude 78°, but reports it "nowhere abundant." Dr. COUES reports it breeding in the higher parts of the Rocky Mountains — "above timber limit." Winters from the Gulf States to Central America.

Anthus spragueii.
SPRAGUE'S PIPIT.

An abundant summer resident of the Great Plains, between Western Manitoba and the base of the Rockies. Winters in Southern Mexico.

Cinclus mexicanus.
AMERICAN DIPPER.

An abundant resident of British Columbia and the eastern slope of the Rockies.

Mimus polyglottos.
MOCKINGBIRD.

One example was taken at Chatham, Ontario, in 1860, by Mr. W. E. SANDYS, and Mr. MCILWRAITH reports that a pair spent the summer of 1883 near Hamilton. Dr. WHEATON reports it breeding in Northern Ohio.

Galeoscoptes carolinensis.
CATBIRD.

This species is a common summer resident from the Maritime Provinces to the Rockies, and is occasionally met with in British Columbia between the Rockies and the Cascades. RICHARDSON found it in the Saskatchewan Valley, in latitude 54°, but on the Atlantic border it has not been found north of latitude 47°.

Harporhynchus rufus.
BROWN THRASHER.

The "Brown Thrush," as it is sometimes called, has not been taken in the Maritime Provinces. Mr. DUNLOP reports it as having become quite common near Montreal during the last ten or fifteen years, though before 1871 it was rare; Dr. HALL, however, recorded it as "common" in his 1839 list. Messrs. WHITE and SCOTT report it as rather common in the vicinity of Ottawa, and it is more or less common in all favorable localities in Ontario as far to the north as Gravenhurst, where Mr. SCRIVEN considers it rare. In Manitoba it is common, or abundant, and ranges to the northern extremity of Lake Winnipeg and west to the eastern base of the Rockies. Winters in the Southern States.

Salpinctes obsoletus.
ROCK WREN.

An uncommon summer resident of British Columbia.

Thryothorus bewickii spilurus.
VIGORS WREN.

A fairly common summer resident of British Columbia.

Troglodytes aëdon.
HOUSE WREN.

There is but one record of the occurrence of this species in the Maritime Provinces — that of a pair observed in New Brunswick by Mr. C. F. BATCHELDER — and Mr. NEILSON reports having met with it once, only, near Quebec; but at Montreal it is fairly common, and is numerous in Southern Ontario.

Troglodytes aëdon parkmanii.
PARKMAN'S WREN.

An abundant summer resident of Manitoba, ranging westward to British Columbia, where it is fairly common.

Troglodytes hyemalis.
WINTER WREN.

The Winter Wren is of more or less common occurrence in all favorable localities from Eastern Manitoba to the Atlantic seaboard, spending the summer and breeding throughout this range, excepting in the more southern portions of Ontario, where it occurs as a migrant, or as a winter visitant. The most northerly point from which it has been reported is Lake Misstassini, where Mr. JOSEPH M. MACOUN found it in March, 1885. Mr. COMEAU reports a pair summering at Point des Monts in 1884.

Troglodytes hyemalis pacificus.
WESTERN WINTER WREN.

An abundant resident of British Columbia.

Cistothorus stellaris.
SHORT-BILLED MARSH WREN.

An abundant summer resident of Western Manitoba and the adjacent Prairies. Mr. SAUNDERS reports finding it rather common on the St. Clair Flats, Ontario.

Cistothorus palustris.
LONG-BILLED MARSH WREN.

This species is a common summer resident in Southern Ontario, and, though it has not been observed near Ottawa, is reported by Messrs. DUNLOP and WINTLE as appearing every spring on Nun's Island, off Montreal. It is a rare summer resident of Manitoba, and westward to the Rockies, and is fairly common in British Columbia, north to Kamloops. Winters south to Guatemala.

Certhia familiaris americana.
BROWN CREEPER.

The Creeper is an uncommon summer resident of the Maritime Provinces, and is rare in Manitoba; but in both Quebec and Ontario it is common, and is found in the southern portions of the latter Province during the entire year, though its general breeding ground is farther north. In British Columbia it is fairly common, north to Kamloops.

Sitta carolinensis.
WHITE-BREASTED NUTHATCH.

This species is said to occur throughout the southern portions of the Dominion, from the Atlantic to the Rockies, but I have not seen any records of it having been taken west of the Georgian Bay. It is sedentary in Southern Ontario, and quite common, and in Quebec and the Maritime Provinces it is a fairly common summer resident.

Sitta carolinensis aculeata.
SLENDER-BILLED NUTHATCH.

A summer resident from Manitoba to the Pacific, being most abundant on the western slopes of the Rocky Mountains and the eastern slopes of the Cascades.

P

Sitta canadensis.
RED-BREASTED NUTHATCH.

Occurs from the Atlantic to the Pacific, and north to Point des Monts and Lake Winnipeg. It is found during the entire year in the northern portions of its range, but in winter only in the more southern portions.

Sitta pygmæa.
PYGMY NUTHATCH.

A summer resident of British Columbia; most abundant east of the Cascades, especially in the Similkameen Valley.

Parus atricapillus.
CHICKADEE.

An abundant resident from the Atlantic border to Lake Huron, and north to about the 50th parallel.

Parus atricapillus septentrionalis.
LONG-TAILED CHICKADEE.

An abundant resident of Manitoba and across the Great Plains to the Rockies.

Parus atricapillus occidentalis.
OREGON CHICKADEE.

An abundant resident along the Pacific coast.

Parus gambeli.
MOUNTAIN CHICKADEE.

This species is common on the eastern slopes of the Cascade Mountains (*Fannin*).

Parus hudsonicus.
HUDSONIAN CHICKADEE.

This species is a common resident of the Maritime Provinces and the islands in the Gulf of St. Lawrence, though in the southern parts of New Brunswick and Nova Scotia it is most numerous in winter. It is abundant during the entire year in Labrador, in Quebec to about latitude 50°, and in the Muskoka district of Ontario. It is rare near the City of Quebec and in the vicinity of Montreal, and for Southern Ontario the only record is of one example taken near Ottawa by Mr. W. L. Scott. Prof. Macoun met it at Porcupine Mountain, and it has been taken at Great Slave Lake, but there is no record for Manitoba or the Hudson's Bay region.

Parus rufescens.
CHESTNUT-BACKED CHICKADEE.

An abundant resident of British Columbia.

Psaltriparus minimus.
BUSH-TIT.

A fairly common bird on Vancouver Island and along the Pacific coast.

Regulus satrapa.
GOLDEN-CROWNED KINGLET.

A common bird from the Atlantic border to the Great Plains, and north to the lower Fur Countries. Breeds chiefly northward of latitude 45°, and winters along the southern border and south to Guatemala.

Regulus satrapa olivaceus.
WESTERN GOLDEN-CROWNED KINGLET.

An abundant summer resident of British Columbia.

Regulus calendula.
RUBY-CROWNED KINGLET.

Occurs throughout Canada, from the United States boundary to the Fur Countries; breeding from about latitude 45° northward. It is abundant in British Columbia, common in the interior, and uncommon or rare along the Atlantic border. Winters south to Guatemala.

Polioptila cærulea.
BLUE-GRAY GNATCATCHER.

A regular and not very rare summer resident of Southern Ontario.

Myadestes townsendii.
TOWNSEND'S SOLITAIRE.

A very rare summer resident of British Columbia (*Fannin*).

Turdus mustelinus.
WOOD THRUSH.

A common summer resident of Southern Ontario, ranging north to Gravenhurst (*Scriven*), and east to Ottawa (*Scott*), and the eastern townships of the Province of Quebec (*Dunlop*). Winters south to Guatemala and Cuba.

Turdus fuscescens.
WILSON'S THRUSH.

This species has been called the "Tawny Thrush" and "Veery." It occurs from the Atlantic border to the Great Plains, and north to Anticosti, breeding throughout.

Turdus fuscescens salicicolus.
WILLOW THRUSH.

Examples of this variety of Wilson's Thrush were taken by Dr. COUES on the Souris Plains, near the 49th parallel.

Turdus aliciæ.
GRAY-CHEEKED THRUSH.

Occurs throughout Canada, north to the Arctics; breeding in high latitudes, and migrating across the southern border.

Turdus aliciæ bicknelli.
BICKNELL'S THRUSH.

The Rev. J. H. LANGILLE reports finding this variety of the Gray-Cheeked Thrush abundant on Mud Island and Seal Island, off the coast of Nova Scotia.

Turdus ustulatus.
RUSSET-BACKED THRUSH.

An abundant summer resident along the Pacific coast, occurring, also, in the interior of British Columbia. Winters in Guatemala.

Turdus ustulatus swainsonii.
OLIVE-BACKED THRUSH.

Occurs from the Atlantic border to the eastern base of the Rockies, and north to the northern shore of the Gulf of St. Lawrence and to the Great Slave Lake region; rather rare between Montreal and Lake Huron, but common to abundant elsewhere. Breeds from about latitude 45° northward.

Turdus aonalaschkæ.
DWARF HERMIT THRUSH.
An abundant summer resident of the Pacific coast, occurring, also, in the interior of British Columbia.

Turdus aonalaschkæ pallasii.
HERMIT THRUSH.
A more or less common bird from the Atlantic border to the Rockies, and north to Anticosti and Lake Misstassini; breeding from about latitude 45°. Winters from the Northern States southward.

Merula migratoria.
AMERICAN ROBIN.
An abundant summer resident throughout Canada, north to the Arctics; breeding throughout. Winters from about latitude 46° (sparingly) southward.

Merula migratoria propinqua.
WESTERN ROBIN.
Occurs abundantly in British Columbia.

Hesperocichla nævia.
VARIED THRUSH.
This species is sedentary along the coast of British Columbia, and is found in the interior during the summer.

Saxicola œnanthe.
WHEATEAR.

Occurs along the Atlantic coast, from Grand Manan to the Arctics. Breeds in Cumberland Bay and south to (probably) the Gulf of St. Lawrence.

Sialia sialis.
BLUEBIRD.

Occurs from the Gulf of St. Lawrence to the eastern edge of the Great Plains, and north to Point des Monts and Lake Winnipeg. It is abundant in Southern Ontario, and common near Montreal, but rare or casual elsewhere. Winters south to Cuba.

Sialia mexicana.
WESTERN BLUEBIRD.

An abundant summer resident of British Columbia.

Sialia arctica.
MOUNTAIN BLUEBIRD.

Occurs from the Great Plains to the Pacific; is abundant in British Columbia, from the Cascade Mountains to the Rockies.

APPENDIX.

APPENDIX.

Since the *A. O. U. Check-List* was issued, several alterations in the nomenclature of North American birds, as well as additions to the avi-fauna, have been made. The following are the changes which affect the present Catalogue:

Somateria mollissima borealis.
GREENLAND EIDER.

This replaces the Eider (*S. Mollissima*) of the *Check-List*.

In a note in Ridgway's *Manual of North American Birds*, the author states that the American bird proves to be not the true *S. mollissima* (which is found in Northern Europe), but a fairly distinguishable race, which he describes and names as above.

Symphemia semipalmata inornata.
WESTERN WILLET.

This is an additional sub-species.

In *The Auk* for April, 1887, Mr. William Brewster separates the western bird from that found along the Atlantic coast, giving the present form sub-specific rank. It is said to differ from typical *semipalmata* in size, color, and markings. Its habitat is given as: "Interior of North America, between the Mississippi and the Rocky Mountains; wintering along the coasts of the Gulf and the Southern Atlantic States (Florida, Georgia, South Carolina)."

I cannot give the distribution in Canada of the two forms, though it is probable that true *semipalmata* ranges from the Atlantic to Manitoba, while *inornata* is restricted to the Prairies.

Scotiaptex cinereum.

GREAT GRAY OWL.

In the *Manual of North American Birds* the name of the genus in which this species is placed is given as above, it being the opinion of the author of the *Manual* that the substitution of *Ulula* for *Scotiaptex*, as the name for this genus, was an error.

Empidonax dificilis.

WESTERN FLYCATCHER.

"Called 'Baird's Flycatcher' in the *A. O. U. Check-List*, but this name belongs properly to *E. bairdii*"—(RIDGWAY).

Corvus corax sinuatus.

MEXICAN RAVEN.

Corvus corax principalis.

NORTHERN RAVEN.

This last is a new sub-species, described by Mr. RIDGWAY in the *Manual*.

The name *sinuatus*, which was given to the American Raven in the *A. O. U. Check-List*, is retained by Mr. RIDGWAY for the Mexican Raven—a form that is found in the "Western United States, and south to Guatemala." From lack of specimens the describer is unable to determine which sub-species the birds of the Eastern United States belong to.

Whether both varieties are represented in the fauna of Canada is yet to be decided, and, until evidence to the contrary is forthcoming, I propose to replace *sinuatus* by *principalis*, allowing the former to remain among the hypothetical cases.

Corvus americanus hesperis.
CALIFORNIA CROW.

This is a new sub-species, which is described in the *Manual of North American Birds.*

The author says of it that it is "a very strongly characterized race, which also differs markedly in habits and notes from the eastern bird." He gives its habitat as "Western United States, north to Washington Territory (Puget Sound), Idaho, Montana, etc., south to Northern Mexico, east to Rocky Mountains."

The California Crow will undoubtedly be found in British Columbia.

Pinicola enucleator canadensis.
AMERICAN PINE GROSBEAK.

This replaces *pinicola enucleator* — Pine Grosbeak — of the *Check-List*, the European bird being the true *enucleator.*

Pinicola enucleator kodiaka.
KODIAK PINE GROSBEAK.

This is a new sub-species.

Mr. RIDGWAY, the describer, in his *Manual*, gives the habitat as "Kodiak to Sitka, Alaska. (Also probably southward to higher Sierra Nevada of California.)"

If this bird ranges to California, it of course occurs in British Columbia in the migrations, and it may possibly be resident in that Province.

Ammodramus caudacutus subvirgatus.
ACADIAN SHARP-TAILED SPARROW.

This is a new variety, discovered by Mr. JONATHAN DWIGHT, Junior, and replaces in our fauna the typical *caudacutus*, which latter has a more southern range.

Mr. DWIGHT describes the present form in *The Auk* for July, 1887.

Dendroica æstiva morcomi.
WESTERN YELLOW WARBLER.

A new sub-species, separated from typical *æstiva*, and described by Mr. H. K. COALE, in *Bulletin of the Ridgway Ornithological Club*, No. 2, April, 1887.

The author characterizes it as "similar to *dendroica æstiva*. Colours lighter. Bill more slender." He considers the race as "so different that it can readily be distinguished from the eastern at sight." The type was taken at Fort Bridger, Utah.

This sub-species probably occurs in Canada, from Manitoba westward.

INDEX.

INDEX.

	PAGE.
Acanthis hornemannii,	82
hornemannii exilipes,	82
linaria,	82
linaria holbœllii,	82
linaria rostrata,	82
Accipiter atricapillus,	55
atricapillus striatulus,	55
cooperi,	55
velox,	55
Actitis macularia,	44
Æchmophorus occidentalis,	1
Ægialitis dubia,	47
hiaticula,	46
meloda,	47
montana,	47
nivosa,	47
semipalmata,	46
vocifera,	46
wilsonia,	47
Agelaius gubernator,	78
phœniceus,	78
Aix sponsa,	22
Albatross, Black-footed,	14
Short-tailed,	14
Sooty,	14
Yellow-nosed,	14
Alca torda,	6
Alle alle,	6
Ammodramus bairdii,	85
caudactus,	86
caudacutus subvirgatus,	125
henslowii,	86
lecontii,	86
princeps,	85

	PAGE.
Ammodramus sandwichensis,	85
sandwichensis alaudinus,	85
sandwichensis savanna,	85
savannarum passerinus,	85
Ampelus cedrorum,	96
garrulus,	95
Anas americana,	21
boschas,	20
carolinensis,	21
crecca,	21
cyanoptera,	21
discors,	21
obscura,	20
penelope,	20
strepera,	20
Anser albifrons gambeli,	28
Anthus pensylvanicus,	109
spragueii,	110
Anstrotomus vociferus,	68
Aphriza virgata,	48
Aquila chrysaëtos,	57
Archibuteo ferrugineus,	57
lagopus sancti-johannis,	57
Ardea candidissima,	32
cœrulea,	32
egretta,	32
herodias,	31
virescens,	32
Arenaria interpres,	48
melanocephela,	48
Asio accipitrinus,	61
wilsonianus,	60
Auk, Great,	6
Razor-billed,	6

INDEX.

	PAGE.
Auklet, Cassin's,	4
Rhinoceros,	4
Avocet, American,	36
Aythya affinis,	22
americana,	22
collaris,	22
marila nearctica,	22
vallisneria,	22
Baldpate,	21
Bartramia longicauda,	43
Bird, Red-billed Tropic,	17
Surf,	48
Yellow-billed Tropic,	16
Bittern, American,	31
Least,	31
Blackbird, Bicolored,	78
Brewer's,	79
Red-winged,	78
Rusty,	79
Yellow-headed,	78
Bluebird,	119
Mountain,	119
Western,	119
Bobolink,	77
Western,	77
Bob-white,	49
Bonasa umbellus togata,	50
umbellus sabini,	51
umbellus umbelloides,	50
Botaurus exilis,	31
lentiginosus,	31
Brachyramphus marmoratus,	4
Brant,	29
Black,	29
Branta bernicla,	29
canadensis,	28
canadensis hutchinsii,	29
canadensis minima,	29
canadensis occidentalis,	29
leucopsis,	30

	PAGE.
Branta nigricans,	30
Bubo virginianus,	62
virginianus arcticus,	62
virginianus saturatus,	62
virginianus subarcticus,	62
Buffle-head,	24
Bunting, Indigo,	92
Lark,	93
Lazuli,	92
Bush-Tit,	115
Buteo borealis,	56
borealis calurus,	56
Itissimus,	57
lineatus,	56
swainsoni,	56
Calamospiza melanocorys,	93
Calcarius lapponicus,	83
ornatus,	84
pictus,	84
Calidris arenaria,	41
Callipepla californica vallicola,	49
Camptolaimus labradorius,	25
Canvas-back,	23
Cardinal,	92
Cardinalis cardinalis,	92
Carpodacus cassini,	81
purpureus,	80
purpureus californicus,	81
Catbird,	110
Catharista atrata,	54
Cathartes aura,	54
Centrocercus urophasianus,	53
Ceophlœus pileatus,	66
Cepphus columba,	5
grylle,	4
mandtii,	5
Cerorhinca monocerata,	4
Certhia familiaris americana,	113
Ceryle alcyon,	64
Chætura pelagica,	68

INDEX. 131

	Page.
Chætura vauxii,	69
Charadrius dominicus,	46
squatorola,	45
Charitonetta albeola,	24
Chat, Yellow-breasted,	108
Chelidon erythrogaster,	94
Chen hyperborea,	27
hyperborea nivalis,	28
rossii,	28
Chickadee,	114
Chestnut-backed,	115
Hudsonian,	115
Long-tailed,	114
Mountain,	114
Oregon,	114
Chondestes grammacus strigatus,	86
Chordeiles virginianus,	68
virginianus henryi,	68
Cinclus mexicanus,	110
Circus hudsonius,	55
Cistothorus palustris,	112
stellaris,	112
Clangula hyemalis,	24
Clivicola riparia,	95
Coccothraustes vespertina,	80
Coccyzus americanus,	63
erythropthalmus,	63
Colaptes auratus,	67
cafer,	67
cafer saturatior,	67
chrysoides,	67
Colinus virginianus,	49
Columba fasciata,	53
Columbigallina passerina,	54
Colymbus auritus,	1
holbœlii,	1
nigricollis californicus,	2
Compsothlypis americana,	100
Contopus borealis,	71

	Page.
Contopus richardsonii,	71
virens,	71
Coot, American,	35
Cormorant,	17
Baird's,	18
Brandt's,	18
Double-crested,	17
Florida,	17
Violet-green,	18
White-crested,	18
Corvus americanus,	76
americanus hesperis,	124
caurinus,	77
corax sinuatus,	76-124
corax principalis,	124
Cowbird,	77
Crake, Corn,	34
Crane, Little Brown,	33
Sandhill,	33
Whooping,	33
Creeper, Brown,	113
Crex crex,	34
Crossbill, American,	81
White-winged,	81
Crow, American,	76
California,	125
Northwest,	76
Crymophilus fulicarius,	35
Cuckoo, Black-billed,	63
Yellow-billed,	63
Curlew, Eskimo,	45
Hudsonian,	45
Long-billed,	44
Cyanocephalus cyanocephalus,	77
Cyanocitta cristata,	75
stelleri,	75
stelleri macrolopha,	75
Cypseloides niger,	68
Dafila acuta,	22
Dendragapus canadensis,	49

Index

	Page.
Dendragapus franklinii,	49
obscurus fuliginosus,	49
obscurus richardsonii,	49
Dendroica æstiva,	101
æstiva morcomi,	126
auduboni,	102
blackburniæ,	103
cærulea,	103
cærulescens,	101
castanea,	103
coronata,	101
maculosa,	102
nigrescens,	104
occidentalis,	104
palmarum,	105
palmarum hypochrysea,	105
pensylvanica,	103
striata,	103
tigrina,	100
townsendi,	104
vigorsii,	104
virens,	104
Dickcissel,	93
Diomedea albatrus,	14
nigripes,	14
Dipper, American,	110
Dolichonyx oryzivorus,	77
oryzivorus albinucha,	77
Dove, Ground,	54
Mourning,	54
Dovekie,	6
Dowitcher,	37
Long-billed,	38
Dryobates pubescens,	64
pubescens gairdnerii,	64
villosus harrisii,	64
villosus leucomelas,	64
Duck, American Scaup,	23
Black,	20
Harlequin,	25

	Page.
Duck, Labrador,	25
Lesser Scaup,	23
Ring-necked,	23
Ruddy,	27
Rufous-crested,	22
Steller's,	25
Wood,	22
Dunlin,	40
Eagle, Bald,	57
Golden,	57
Ectopistes migratorius,	53
Egret, American,	32
Eider,	25
American,	26
Greenland,	123
King,	25
Pacific,	25
Elanoides forficatus,	55
Empidonax acadicus,	72
difficilis,	72–124
flaviventris,	72
hammondi,	73
minimus,	73
obscurus,	73
pusillus,	72
pusillus traillii,	73
Eniconetta stelleri,	25
Ereunetes occidentalis,	41
pusillus,	41
Erismatura rubida,	27
Falco columbarius,	59
columbarius suckleyi,	59
islandus,	58
peregrinus anatum,	59
peregrinus pealei,	59
richardsonii,	60
rusticolus,	58
rusticolus gyrfalco,	58
rusticolus obsoletus,	59
sparverius,	60

INDEX.

	PAGE.
Falcon, Peale's,	59
Finch, California Purple,	81
Cassin's Purple,	81
Purple,	80
Flicker,	67
Gilded,	67
Northwestern,	67
Red-shafted,	67
Flycatcher, Acadian,	72
Baird's,	72
Crested,	71
Hammond's,	73
Least,	73
Little,	72
Olive-sided,	71
Scissor-tailed,	70
Traill's,	73
Western,	124
Wright's,	73
Yellow-bellied,	72
Fratercula arctica,	3
arctica glacialis,	3
corniculata,	4
Fregata aquila,	19
Fulica americana,	35
Fulmar,	15
Lesser,	15
Pacific,	15
Fulmaris glacialis,	15
glacialis glupischa,	15
glacialis minor,	15
Gadwall,	20
Galeoscoptes carolinensis,	110
Gallinago delicata,	37
gallinago,	37
Gallinula galeata,	35
Gallinule, Florida,	35
Purple,	35
Gannet,	17
Gavia alba,	7
Gelochelidon nilotica,	14
Geothlypis agilis,	107
formosa,	106
macgillivrayi,	10
philadelphia,	107
trichas,	108
trichas occidentalis,	108
Glaucidum gnoma,	63
Glaucionetta clangula americana,	24
islandica,	24
Gnatcatcher, Blue-grey,	116
Godwit, Hudsonian,	42
Marbled,	41
Pacific,	41
Golden-eye, American,	24
Barrow's,	24
Goldfinch, American,	83
Goose, Amer. White-fronted,	28
Barnacle,	29
Cackling,	29
Canada,	28
Greater Snow,	28
Hutchins',	29
Lesser Snow,	27
Ross's Snow,	28
White-cheeked,	29
Goshawk, American,	55
Western,	56
Grackle, Bronzed,	80
Grebe, American Eared,	2
Holbœll's,	1
Horned,	1
Pied-billed,	2
Western,	1
Grosbeak, Black-headed,	92
American Pine,	125
Blue,	92
Evening,	80
Kodiak Pine,	125

	PAGE.
Grosbeak, Pine,	80
Rose-breasted,	92
Grouse, Canada,	50
Canadian Ruffed,	50
Columbian Sharp-tailed,	52
Franklin's,	50
Gray Ruffed,	50
Oregon Ruffed,	51
Prairie Sharp-tailed,	53
Richardson's,	49
Sage,	53
Sharp-tailed,	52
Sooty,	49
Grus americana,	33
canadensis,	33
mexicana,	33
Guillemot, Black,	4
Mandt's,	5
Pigeon,	5
Guiraca cærulea,	92
Gull, American Herring,	9
Bonaparte's,	11
California,	10
Franklin's,	11
Glacous,	8
Glacous-winged,	8
Great Black-backed,	9
Herring,	9
Heermann's,	10
Iceland,	8
Ivory,	7
Kumlien's,	8
Laughing,	10
Mew,	10
Pallas's,	9
Ring-billed,	10
Ross's,	11
Sabine's,	11
Short-billed,	10
Western,	9
Gyrfalcon,	58
Black,	59
Gray,	58
White,	58
Habia ludoviciana,	92
melanocephala,	92
Hæmatopus bachmani,	49
palliatus,	48
Haliæetus leucocephalus,	58
Harporhynchus rufus,	111
Hawk, American Rough-leg'd,	57
American Sparrow,	60
Broad-winged,	57
Cooper's,	55
Duck,	59
Ferruginous Rough-leg'd,	57
Marsh,	55
Pigeon,	59
Red-shouldered,	56
Red-tailed,	56
Sharp-shinned,	55
Swainson's,	56
Western Red-tailed,	56
Helminthophila celata,	99
celata lutescens,	99
chrysoptera,	98
peregrina,	99
ruficapilla,	98
Hen, Prairie,	52
Heron, Black-crowned Night,	33
Great Blue,	31
Green,	33
Little Blue,	32
Snowy,	32
Hesperocichla, nævia,	118
Heteractitis incanus,	43
Himantopus mexicanus,	36
Histrionicus histrionicus,	25
Hummingbird, Allen's,	69
Black-chinned,	69

INDEX.

	Page.
Hummingbird, Calliope,	70
Ruby-throated,	69
Rufous,	69
Hydrochelidon leucoptera,	14
nigra surinamensis,	13
Ibis, Glossy,	31
Icteria virens,	108
Icterus bullocki,	79
galbula,	79
spurius,	79
Ionorius martinica,	35
Jaeger, Parasitic,	7
Pomarine,	6
Long-tailed,	7
Jay, Alaskan,	76
Blue,	75
Canada,	75
Labrador,	76
Long-crested,	75
Oregon,	76
Pinon,	77
Rocky Mountain,	76
Steller's,	75
Junco hyemalis,	89
hyemalis oregonus,	89
Junco, Oregon,	89
Slate-colored,	89
Killdeer,	46
Kingbird,	70
Arkansas,	70
Kingfisher, Belted,	64
Kinglet, Golden-crowned,	115
Ruby-crowned,	116
Western Golden-crowned,	115
Kite, Swallow-tailed,	55
Kittawake,	7
Pacific,	8
Knot,	38
Lagopus lagopus,	51
leucurus,	52

	Page.
Lagopus rupestris,	51
rupestris reinhardti,	52
Lanius borealis,	96
ludovicianus excubitorides,	96
Lark, Desert Horned,	74
Horned,	74
Pallid Horned,	74
Prairie Horned,	74
Streaked Horned,	74
Larus argentatus,	9
argentatus smithsonianus,	9
articilla,	10
brachyrhynchus,	10
cachinnans,	9
californicus,	10
canus,	10
delawarensis,	10
franklinii,	10
glaucescens,	8
glaucus,	8
heermanni,	10
kumlieni,	8
leucopterus,	8
marinus,	9
occidentalis,	9
philadelphia,	10
Leucosticte tephrocotis,	81
tephrocotis littoralis,	82
Leucosticte, Gray-crowned,	81
Hepburn's,	82
Limosa fedoa,	41
hæmastica,	42
lapponica baueri,	41
Longspur, Chestnut-collared,	84
Lapland,	83
McCown's,	84
Smith's,	84
Loon,	2
Black-throated,	2
Pacific,	3

INDEX.

	PAGE.
Loon, Red-throated,	3
Yellow-billed,	2
Lophodytes cucullatus,	19
Loxia curvirostra minor,	81
leucoptera,	81
Lunda cirrhata,	3
Macrorhamphus griseus,	37
scolopaceus,	38
Magpie, American,	75
Mallard,	20
Man-o'-War Bird,	19
Martin, Purple,	94
Meadowlark,	78
Western,	78
Megalestris skua,	6
Megascops asio,	61
asio kennicottii,	62
Melanerpes carolinus,	66
erythrocephalus,	66
torquatus,	66
Meleagris gallopavo,	53
Melospiza fasciata,	89
fasciata guttata,	90
fasciata rufina,	90
georgiana,	90
lincolni,	90
Merganser americanus,	19
serrator,	19
Merganser, American,	19
Hooded,	19
Red-breasted,	19
Merlin, Black,	59
Richardson's,	60
Merula migratoria,	118
migratoria propinqua,	118
Micropalama himantopus,	38
Milvulus forficatus,	70
Mimus polyglottos,	110
Mniotilta varia,	98
Mockingbird,	110

	PAGE.
Molothrus ater,	77
Murre,	5
Brünnich's,	5
California,	5
Pallas's,	5
Murrelet, Marbled,	4
Temninck's,	4
Myadestes townsendii,	116
Myiarchus crinitus,	71
Netta rufina,	22
Nighthawk,	68
Western,	68
Numenius borealis,	45
hudsonicus,	45
longirostris,	44
Nutcracker, Clarke's,	77
Nuthatch, Pygmy,	114
Red-breasted,	114
Slender-billed,	113
White-breasted,	113
Nyctala acadica,	61
tengmalmi richardsoni,	61
Nyctea nyctea,	62
Nycticorax nycticorax nævius,	33
Oceanites oceanicus,	16
Oceanodroma furcata,	16
leucorhoa,	16
Oidemia americana,	26
deglandi,	27
perspicillata,	27
Old-squaw,	21
Olor buccinator,	30
columbianus,	30
Oreortyx pictus,	49
Oriole, Baltimore,	79
Bullock's,	79
Orchard,	79
Osprey, American,	60
Otocoris alpestris,	74
alpestris arenicola,	74

Index.

	Page.
Otocoris alpestris leucolæma,	74
alpestris praticola,	74
alpestris strigata,	74
Oven-bird,	105
Owl, American Barn,	60
American Hawk,	63
American Long-eared,	60
Arctic Horned,	62
Barred,	61
Burrowing,	63
Dusky Horned,	62
Great Gray,	61–124
Great Horned,	62
Kennicott's Screech,	62
Pygmy,	63
Richardson's,	61
Saw-whet,	61
Screech,	61
Short-eared,	61
Snowy,	62
Western Horned,	62·
Oyster-catcher, American,	48
Black,	49
Pandion haliaëtus carolinensis,	60
Partridge, Mountain,	49
Valley,	49
Parus atricapillus,	114
atricapillus occidentalis,	114
atricapillus septentrionalis,	114
gambeli,	114
hudsonicus,	115
rufescens,	115
Passer domesticus,	83
Passerella iliaca,	90
iliaca unalaschensis,	91
Passerina amœna,	93
cyanea,	92
Pavoncella pugnax,	43
Pediocætes phasianellus,	52
phasianellus campestris,	53

	Page.
Pediocætes phasianellus columbianus,	52
Pelecanus californicus,	19
erythrorhynchos,	18
Pelican, American White,	18
California Brown,	19
Perisoreus canadensis,	75
canadensis capitalis,	76
canadensis fumifrons,	76
canadensis nigricapillus,	76
obscurus,	76
Petrel, Fork-tailed,	16
Leach's,	16
Stormy,	16
Wilson's,	16
Petrochelidon lunifrons,	94
Peucæa ruficeps,	89
Pewee, Western Wood,	72
Wood,	71
Phaëthon æthereus,	17
flavirostris,	16
Phalacrocorax carbo,	17
dilophus,	17
dilophus cincinatus,	18
dilophus floridanus,	17
pelagicus resplendens,	18
pelagicus robustus,	18
penicillatus,	18
Phalarope, Northern,	35
Red,	35
Wilson's,	36
Phalaropus lobatus,	35
tricolor,	36
Philohela minor,	37
Phœbe,	71
Say's,	71
Phœbetria fuliginosa,	14
Pica pica hudsonica,	75
Picoides americanus,	65
arcticus,	65

138 INDEX.

	PAGE		PAGE
Picicorvus columbianus,	77	Ptarmigan, Rock,	51
Pigeon, Band-tailed,	53	White-tailed,	.52
Passenger,	53	Willow,	51
Pinicola enucleator,	80	Ptychoramphus aleuticus,	4
enucleator canadensis,	125	Puffin,	3
enucleator kodiaka,	125	Horned,	4
Pintail,	22	Large-billed,	3
Pipilo erythrophthalmus,	91	Tufted,	3
maculatus arcticus,	91	Puffinus major,	15
maculatus oregonus,	91	puffinus,	15
Pipit, American,	109	stricklandi,	15
Sprague's,	110	Quiscalus quiscula æneus,	80
Piranga erythromelas,	93	Rail, King,	34
ludoviciana,	93	Virginia,	34
rubra,	93	Yellow,	34
Plautus impennis,	6	Rallus elegans,	34
Plectrophenax nivalis,	83	virginianus,	34
Plegadis autumnalis,	31	Raven, American,	76
Plover, American Golden,	46	Mexican,	124
Black-bellied,	45	Northern,	124
Little Ring,	47	Recurvirostra americana,	36
Mountain,	48	Redhead,	22
Piping,	47	Redpoll,	82
Ring,	46	Greenland,	82
Semipalmated,	46	Greater,	82
Snowy,	47	Hoary,	82
Wilson's,	47	Holbœll's,	82
Podilymbus podiceps,	2	Redstart, American,	107
Polioptila cærulea,	116	Red-tail, Western,	56
Poocætes, gramineus,	84	Regulus calendula,	116
gramineus confinis,	84	satrapa,	115
Porzana carolina,	34	satrapa olivaceus,	115
noveboracensis,	34	Rhodostethia rosea,	11
Prairie Hen,	52	Rissa tridactyla,	7
Procellaria pelagica,	16	tridactyla pollicaris,	8
Progne subis,	94	Robin, American,	118
Protonotaria citera,	98	Western,	118
Psaltriparus minimus,	105	Rough leg, Ferruginous,	57
Pseudogryphus californianus,	54	Ruff,	43
Ptarmigan, Reinhardt's,	52	Rynchophanes mccownii,	84

	PAGE
Rynchops nigra,	14
Salpinctes obsoletus,	111
Sanderling,	41
Sandpiper, Baird's,	39
Bartramian,	43
Buff-breasted,	44
Curlew,	40
Green,	42
Least,	40
Pectoral,	39
Purple,	39
Red-backed,	40
Semipalmated,	41
Solitary,	42
Spotted,	44
Stilt,	38
Western,	41
White-rumped,	39
Sapsucker, Red-naped,	66
Red-breasted,	66
Yellow-bellied,	65
Saxicola œnanthe,	119
Sayornis phœbe,	71
saya,	71
Scolecophagus carolinus,	79
cyanocephalus,	79
Scolopax rusticola,	37
Scoter, American,	26
Surf,	27
White-winged,	27
Scotiaptex cinereum,	124
Seiurus aurocapillus,	105
motacilla,	106
noveboracensis,	105
noveboracensis notabilis,	106
Setophaga ruticilla,	109
Shearwater, Greater,	15
Manx,	15
Shearwater, Sooty,	15
Shoveller,	22

	PAGE
Shrike, Northern,	96
White-rumped,	96
Sialia arctica,	119
mexicana,	119
sialis,	119
Siskin, Pine,	83
Sitta, canadensis,	114
carolinensis,	113
carolinensis aculeata,	113
pygmæa,	113
Skimmer, Black,	14
Skua,	6
Snipe, European,	37
Wilson's,	37
Snowflake,	83
Solitaire, Townsend's,	116
Somateria dresseri,	26
mollissima,	25
mollissima borealis,	123
spectabilis,	26
v-nigra,	26
Sora,	34
Sparrow,	86
Acadian Sharp-tailed,	125
Baird's,	85
Brewer's,	88
Chipping,	88
Clay-colored,	88
European House,	83
Field,	88
Fox,	88
Gambel's,	87
Golden-crowned,	87
Grasshopper,	85
Harris's,	86
Henslow's,	86
Intermediate,	87
Ipswich,	85
Leconte's,	86
Lincoln's,	88

	Page.		Page.
Sparrow, Rufous-crowned,	88	Sterna hirundo,	13
Rusty Song,	88	maxima,	12
Sandwich,	85	paradisæa,	13
Savanna,	85	Sandvicensis acuflavida,	12
Sharp-tailed,	86	tschegrava,	12
Song,	88	Stilt, Black-necked,	36
Sooty Song,	88	Strix pratincola,	60
Swamp,	88	Sturnella magna,	78
Townsend's,	88	magna neglecta,	78
Tree,	88	Sula bassana,	17
Vesper,	84	Surf Bird,	48
Western Chipping,	88	Surnia ulula caparoch,	63
Western Lark,	86	Swallow, Bank,	95
Western Savanna,	85	Barn,	94
Western Tree,	88	Cliff,	94
Western Vesper,	84	Rough-winged,	95
White-crowned,	87	Tree,	94
White-throated,	87	Violet-green,	95
Spatula clypeata,	22	Swan, Trumpeter,	30
Speotyto cunicularia hypogæa,	63	Whistling,	30
Sphyrapicus ruber,	65	Swift, Black,	69
varius,	65	Chimney,	69
varius nuchalis,	65	Vaux's,	69
Spinus pinus,	83	Sylvania canadensis,	109
tristis,	83	mitrata,	108
Spiza americana,	93	pusilla,	108
Spizella breweri,	88	pusilla pileolata,	109
monticola,	88	Symphemia semipalmata,	43
monticola ochracea,	88	semipalmata inornata,	123
pallida,	88	Synthliboramphus wumizus-	
pusilla,	88	ume,	4
socialis,	88	Syrnium nebulosum,	61
socialis arizonæ,	88	Tachycineta bicolor,	94
Stelgidopteryx serripennis,	95	thalassina,	95
Stercorarius longicaudus,	7	Tannager, Louisiana,	93
parasiticus,	7	Scarlet,	93
pomarinus,	6	Summer,	93
Sterna antillarum,	13	Tatler, Wandering,	43
dougalli,	13	Teal, Blue-winged,	21
forsteri,	12	Cinnamon,	21

Index.

	Page.
Teal, European,	21
Green-winged,	21
Tern, Arctic,	13
Black,	13
Cabot's,	12
Caspian,	12
Common,	13
Forster's,	12
Gull-billed,	12
Least,	13
Roseate,	13
Royal,	12
White-winged Black,	14
Thalassogeron culminatus,	14
Thrasher, Brown,	111
Thrush, Bicknell's,	117
Dwarf Hermit,	118
Gray-cheeked,	117
Hermit,	118
Olive-backed,	117
Russet-backed,	117
Varied,	108
Willow,	117
Wilson's,	116
Wood,	116
Thryothorus bewickii spilurus,	111
Totanus flavipes,	42
melanoleucus,	42
ochropus,	42
solitarius,	42
Towhee,	91
Arctic,	91
Oregon,	91
Tringa alpina,	40
alpina pacifica,	40
bairdii,	39
canutus,	38
feruginea,	40
fuscicollis,	39
maculata,	39

	Page.
Tringa maritima,	39
minutilla,	40
Trochilus alexandri,	69
alleni,	69
calliope,	70
colubris,	69
rufus,	69
Troglodytes aëdon,	111
aëdon parkmanii,	112
hyemalis,	112
hyemalis pacificus,	112
Tropic Bird, Red-billed,	17
Yellow-billed,	16
Tryngites subruficollis,	44
Turdus aliciæ,	117
aliciæ bicknelli,	117
aonalaschkæ,	118
aonalaschkæ pallasi,	118
fuscescens,	116
fuscescens salicicolus,	117
mustelinus,	116
ustulatus,	117
ustulatus swainsoni,	117
Turkey, Wild,	53
Turnstone,	48
Black,	48
Tympanuchus americanus,	52
Tyrannus tyrannus,	70
verticalis,	70
Ulula cinerea,	61
Uria lomvia,	5
lomvia arra,	5
troile,	5
troile californica,	5
Urinator adamsii,	2
arcticus,	2
imber,	2
lumme,	3
pacificus,	3
Vireo flavifrons,	97

INDEX.

	PAGE.
Vireo flavoviridis,	96
gilvus,	97
noveboracensis,	98
olivaceus,	96
philadelphicus,	97
solitarius,	97
solitarius cassinii,	97
Vireo, Blue-headed,	97
Cassin's,	97
Philadelphia,	97
Red-eyed,	96
Warbling,	97
White-eyed,	98
Yellow-green,	96
Yellow-throated,	97
Vulture, Black,	54
California,	54
Turkey,	54
Warbler, Audubon's,	102
Bay-breasted,	103
Black and White,	98
Blackburnian,	103
Black-poll,	103
Black-throated Blue,	101
Black-throated Gray,	104
Black-throated Green,	104
Canadian,	109
Cape May,	100
Cerulean,	103
Chestnut-sided,	103
Connecticut,	107
Golden-winged,	98
Hermit,	104
Hooded,	108
Kentucky,	106
Lutescent,	99
Macgillivray's,	107
Magnolia,	102
Mourning,	107
Myrtle,	101

	PAGE.
Warbler, Nashville,	98
Orange-crowned,	99
Palm,	105
Parula,	100
Pileolated,	109
Pine,	104
Prothonotary,	98
Tennessee,	99
Townsend's,	104
Western Yellow,	126
Wilson's,	108
Yellow,	101
Yellow Palm,	105
Water-Thrush,	105
Grinnell's,	106
Louisiana,	106
Waxwing, Bohemian,	95
Cedar,	96
Wheatear,	119
Whip-poor-will,	68
Widgeon,	20
Willet,	43
Western,	123
Woodcock, American,	37
European,	37
Woodpecker, Arctic Three-toed,	65
American Three-toed,	65
Downy,	64
Gairdner's,	64
Harris's,	64
Lewis's,	66
Northern Hairy,	64
Pileated,	66
Red-bellied,	66
Red-headed,	66
White-headed,	64
Wren, House,	111
Long-billed Marsh,	112
Parkman's,	112

Index.

	PAGE
Wren, Rock,	111
Short-billed Marsh,	112
Vigor's,	111
Winter,	112
Western Winter,	112
Xanthocephalus xanthocephalus,	78
Xema sabinii,	11
Xenopicus albolarvatus,	64
Yellow-legs,	42
Yellow-legs, Greater,	42
Yellow-throat, Maryland,	108
Western,	108
Zenaidura macroura,	54
Zonotrichia albicollis,	87
coronata,	87
gambeli,	87
intermedia,	87
leucophrys,	87
querula,	86

www.ingramcontent.com/pod-product-compliance
Lightning Source LLC
Chambersburg PA
CBHW030319170426
43202CB00009B/1068